C0-ALJ-227

VADOSE ZONE MODELING OF ORGANIC POLLUTANTS

Edited by
STEPHEN C. HERN
SUSAN M. MELANCON

Library of Congress Cataloging-in-Publication Data

Vadose zone modeling of organic pollutants.

Bibliography: p.
Includes index.
1. Organic water pollutants. 2. Zone of aeration.
3. Soil chemistry. I. Hern, S. C. II. Melancon, S. M.
(Susan M.)
TD427.07V33 1986 628.5′5 86-20139
ISBN 0-87371-042-8

Second Printing 1987

LEWIS PUBLISHERS, INC.
121 South Main Street, P.O. Drawer 519, Chelsea, Michigan 48118

PRINTED IN THE UNITED STATES OF AMERICA

PREFACE

This book was conceived to address the needs of a growing body of individuals working with soil fate and transport models. A variety of stochastic and deterministic soil leaching models have been developed in the past decade, particularly to measure the transport and transformation of organic pollutants moving through the vadose zone. The U.S. Environmental Protection Agency (U.S. EPA) has supported the development and testing of three such models, largely for screening purposes: the Seasonal Soil Compartment Model (SESOIL), the Pesticide Root Zone Model (PRZM), and the Pesticide Analytical Model (PESTAN). This book seeks to provide the reader with a general overview of the uses and limitations of vadose zone models and with a generic set of guidelines for field collection of data necessary to calibrate and evaluate their predictive capabilities. The book specifically examines the assumptions, data requirements, and processes underlying SESOIL, PRZM, and PESTAN.

These guidelines are intended for use by readers coming from a diversity of scientific backgrounds. However, it is recommended that all users have a working knowledge or at least a bachelor's level training in mathematics, soil science, engineering, modeling, or one of the physical sciences. The first four chapters of the book provide background information to the general model user. Chapter 1 includes an overview of the basic soil processes represented in vadose zone models, including process definitions and common model assumptions. In particular, SESOIL, PRZM, and PESTAN transformation processes presumed operative in the vadose zone are examined more closely. This chapter provides the reader with tables of references for process rate constants and field collection methods for parameters commonly input to vadose zone models. Also in this chapter, the reader is directed to Section II of the book, where detailed process descriptions, mathematical derivations of significant equations underlying these processes, and information regarding the spatial variability of soil properties can be found. The reader is encouraged to be cognizant that although the process mathematical derivations are separated from the main text for ease in reading, an understanding of the material included in Appendix A is essential if vadose zone models are to be meaningfully used for screening purposes or in field validation attempts.

Chapter 3 introduces a stepwise generic approach for the implementation of a data acquisition strategy which can be used in field model testing. This chapter provides guidelines on criteria to be used in model selection and establishment of validation acceptance criteria, as well as site and compound selection, and implementation of a field sampling program. Chapters 4 and 5 then present two scenarios describing actual field validation attempts using the three models we have more closely scrutinized. In the first of these examples (Chapter 4), the generic guidelines described in Chapter 3 are followed step-by-step to demonstrate how the guidelines can aid in the design of a field study intended to yield data for use in subsequent model validations. Thus, this scenario will demonstrate the steps that must be taken to deal with such considerations as site and compound selection, model calibration and sensitivity testing, field data collection, chemical analyses, and quality assurance. The second scenario (Chapter 5) is based on field data which were then applied to various model validation attempts. This scenario is the more common in the environmental modeling literature and illustrates for the reader the difficulties in post hoc comparisons of field data with model predictions when the original study design had a different data collection goal.

This document provides a unique compilation of many of the factors that must be considered in the field testing and application of vadose zone fate and transport models. It does not address models or processes specifically applicable either to the soil saturated zone or to the movement and transformation of inorganic compounds. Furthermore, many multidimensional environmental elements affecting the processes under consideration, such as the effects of surface runoff, plant washoff, or distribution of plume emissions, are beyond the scope of this book. However, references are given throughout the book to assist the user in the appropriate application of vadose zone models. Although these guidelines are directed toward areas impacted by organic pollutants, they contain valuable information for the model user working with other vadose zone pollution problems as well.

We gratefully acknowledge the efforts of all those individuals who helped in the production of this book. The numerous industrial, academic, and governmental reviewers of both the interim and final versions of this document contributed substantially to its organization and technical content. In particular, we wish to acknowledge Dr. Loran G. Wilson at the University of Arizona who met with a panel of the chapter authors, and whose recommendations for the interim document led to the present book's final structure. In addition, we would like to thank Mr. Charles N. Smith at the U.S. EPA, Environmental Research Laboratory, Athens, Georgia, who provided assistance in the preparation of Chapter 4. Finally, we wish to thank Lillian Steele and Susie Reppke of Computer Sciences Corporation for their tireless word processing efforts during the multiple major revisions of this lengthy work.

Stephen C. Hern
Susan M. Melancon

CONTRIBUTORS

Anthony S. Donigian, Jr.
Aqua Terra Consultants
Mountain View, California

Kenneth F. Hedden
Exposure Assessment Research Division
U.S. Environmental Protection Agency
Las Vegas, Nevada

Stephen C. Hern
Exposure Assessment Research Division
U.S. Environmental Protection Agency
Las Vegas, Nevada

William A. Jury
Department of Soil and Environmental Science
University of California – Riverside
Riverside, California

Susan M. Melancon
Environmental Research Center
University of Nevada – Las Vegas
Las Vegas, Nevada

James E. Pollard
Environmental Research Center
University of Nevada – Las Vegas
Las Vegas, Nevada

P. Suresh Chandra Rao
Soil Science Department
University of Florida
Gainesville, Florida

Jerald L. Schnoor
Department of Civil and Environmental Engineering
University of Iowa
Iowa City, Iowa

Richard L. Valentine
Department of Civil and Environmental Engineering
University of Iowa
Iowa City, Iowa

CONTENTS

TABLES

FIGURES

SECTION I

AN OVERVIEW OF CHEMICAL FATE AND TRANSPORT MODELING IN THE VADOSE ZONE

CHAPTER 1

OVERVIEW OF TERRESTRIAL PROCESSES AND MODELING

A. S. Donigian, Jr.
Aqua Terra Consultants, Palo Alto, California

P. S. C. Rao
University of Florida, Gainesville, Florida

INTRODUCTION

The terrestrial environment which extends from the top of the growing vegetation to the capillary fringe of ground water is the primary home for most living things on earth. We deliberately introduce chemicals into this environment to grow and expand our food supply, to protect us and our crops from pests and disease, and to dispose of our wastes; unintended entry also occurs through transport accidents, inaccurate or inappropriate application procedures, and leaking storage facilities. The joint occupancy and use of the terrestrial environment by chemicals and other living organisms (humans, animals, and plants) can lead to significant human and animal exposures to these often toxic chemicals with resulting detrimental health impacts.

Since chemicals are necessary for maintenance of current life styles and standards of living, a better understanding of the fate and migration of chemicals introduced to the terrestrial environment would allow us to better evaluate and control the resulting exposure and health risk. This environment is a dynamic, interdependent system of abiotic and biotic factors that are linked by physical, chemical, and biological processes. Chemicals introduced into this system use these linkages to migrate within and between the various media and are in turn transformed and degraded as they move.

This chapter presents an overview of these processes influencing the fate and migration of chemicals in the terrestrial environment. This overview provides the framework for discussing the underlying concepts of mathematical models that have been developed both as research tools (to help us better understand the terrestrial system) and as regulatory or management tools (to help us assess and control the exposure and risks resulting from chemical use and waste disposal). Model classifications are discussed, selected models are described, model selection criteria and guidelines are presented, and model limitations are explored in terms of their ability to represent chemical fate and movement in the soil environment.

SOIL SYSTEM PROCESSES

Figure 1-1 schematically demonstrates the complex and dynamic interaction of processes controlling the fate and transport of chemicals in the soil environment. In order to provide some order to the multitude of individual processes that can occur, we have grouped the terrestrial processes into the following five categories, as suggested by Rao and Jessup [1] and Wagenet and Rao [2]:

- Transport
- Sorption
- Transformation/Degradation
- Volatilization
- Plant Processes

Although we have listed these categories individually, it is the dynamic, temporal, and spatial interaction among these groups of processes that determines the ultimate disposition of chemicals in the soil. In the paragraphs below, each of these process groups is defined and discussed with specific emphasis on their role in affecting chemical fate and transport and on their mutual interactions. These processes are discussed more thoroughly in Chapter 2.

Transport

After chemicals are introduced into the terrestrial environment, they can move by three separate pathways and mechanisms: by runoff and erosion to the aquatic environment, by volatilization to the air environment, and by lateral or vertical movement (or leaching) to ground water. The transport component is critical to the environmental fate of a chemical; without it, environmental contamination problems would be minimal. Chemicals applied for agricultural or silvicultural purposes would remain on target areas until they were degraded or transformed. Chemical spills would impact only the immediate spill site, and contamination from hazardous waste disposal and land treatment sites would be limited to the facility confines. However, we are well aware that agricultural chemical runoff does occur, that chemical applications and spills can move over the land and through the soil to contaminate surface and ground-water supplies, and that hazardous wastes often tend to migrate away from their storage, disposal, and treatment facilities.

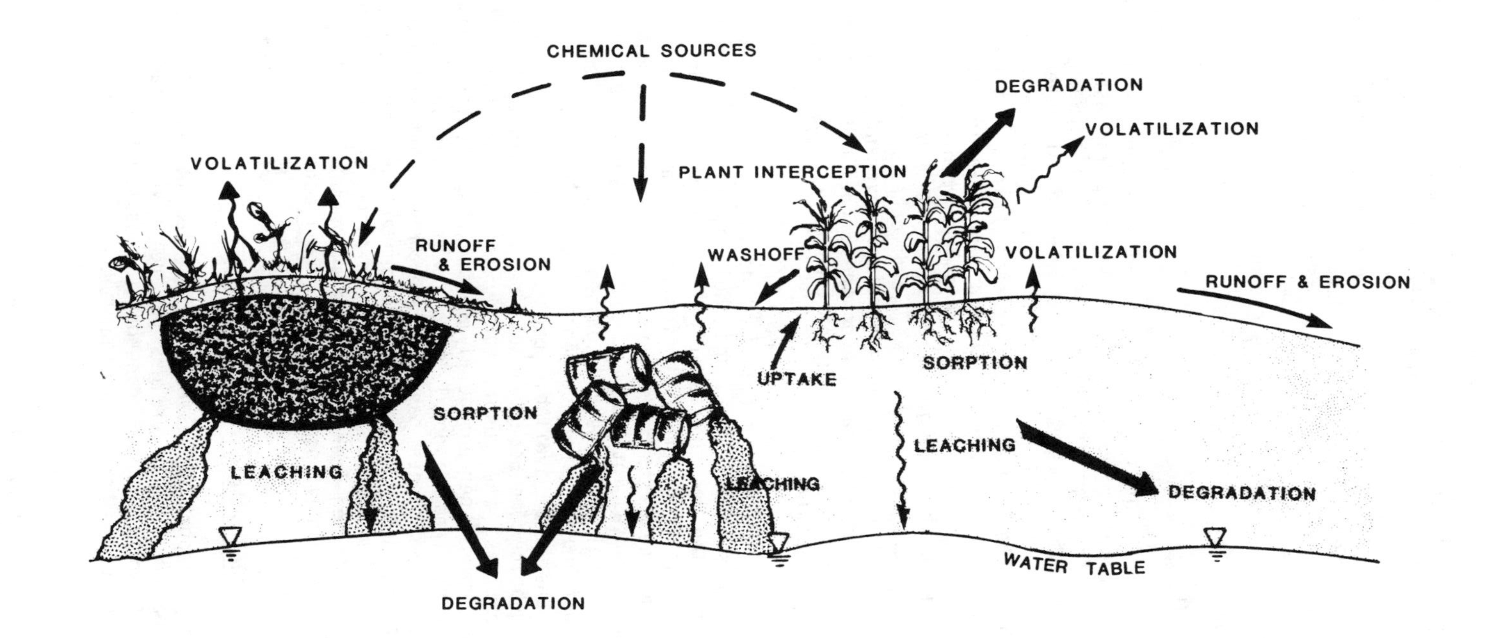

Figure 1-1. Chemicals in the soil environment.

The emphasis in this manual is on the vertical movement of chemicals through the soil to evaluate the potential for ground-water contamination. However, it is important to recognize that the other transport processes can occur depending on specific environmental, soil, chemical, and management conditions and practices. Moreover, the relative significance of runoff/erosion and volatilization will affect the amount of chemical that remains in the soil and that can subsequently leach to ground water.

Sorption

Since movement of chemicals to ground water is primarily a liquid-phase process involving water movement and associated dissolved solute, the partitioning of the chemical between the sorbed and dissolved phases is a critical factor in determining how rapidly the chemical will leach. Chemicals that sorb readily to soil organic matter and clay particles will not migrate significantly away from the region of the soil profile in which they are initially placed or applied. Thus, highly-sorbed chemicals applied on soils will usually remain on or near the land surface unless extensive soil cracks, macropores, and other preferential flow paths exist to allow for a rapid vertical movement of soil particles and associated chemicals. For these chemicals, transport by surface runoff and erosion may be of greater environmental significance than chemical leaching.

On the other hand, chemicals that do not sorb will exist primarily in the dissolved phase, and their movement to ground water will be controlled by the relative timing and amount of water applications (either by precipitation or irrigation), soil characteristics, and the "fate" processes discussed below. Many chemicals of environmental interest are moderately sorbed and thus exist in both the dissolved and sorbed phases. For these chemicals, the surface-runoff and erosion components usually comprise a small fraction of the total available chemical in the soil although the resulting aquatic concentrations can be significant. For agricultural applications, the runoff and erosion losses for pesticides are usually a small percentage of the total application amount [3, 4,5]. However, land disposal and treatment options for highly sorbed waste constituents can generate significant surface runoff loadings if not properly designed to prevent or minimize surface runoff. Thus, for chemicals in the mid-range of sorption properties, both runoff and leaching may be important, depending on the specific chemical, environmental, soil, and management conditions.

Transformation/Degradation

Whereas transport processes specify the vehicles for chemical movement, and sorption determines the relative importance of dissolved versus sorbed phase transport, transformation and degradation processes ultimately determine whether a chemical will persist long enough in the soil environment to be of concern. We have lumped the transformation and degradation processes to include the primary chemical and biological mechanisms that encompass the "fate" of a chemical and determine its persistence in the soil. The key processes include biotransformation, chemical hydrolysis, photolysis, and oxidation-reduction. These individual processes are defined and summarized in Chapter 2 and discussed in detail in Chapters 6-11. Although volatilization and plant processes can be considered as "fate" processes, they also include transport considerations and other special characteristics that require individual discussions (below).

Non-persistent chemicals (i.e., half lives less than 15-20 days) that remain in the soil at significant levels for only a few months after initial placement or application are highly dependent on the relative timing of rainfall or irrigation events to demonstrate significant mobility and potential for contamination. Thus, for these chemicals, the transport vehicles of surface runoff and soil water movement must occur during the period when the chemical is resident in the soil profile in order for chemical transport to occur. The classic example of this is agricultural runoff of relatively non-persistent pesticides. Numerous studies have shown that the highest pesticide runoff concentrations consistently occur in the first few runoff-producing storm events following a field application [4,6,7,8,9].

A somewhat analogous but different situation occurs for chemical leaching; the highest concentrations with the deepest penetration into the soil profile will likely occur during the few months immediately following placement or application. However, superimposed on this vertical transport is a change in the specific transformation/degradation processes and their rates, that may be active at different depths of the soil profile. Processes such as photolysis and volatilization occur primarily in the top few centimeters of the soil profile, whereas biotransformation, hydrolysis, and oxidation-reduction can occur throughout the crop root zone and below. Although the specific rates at which these processes occur depend on individual chemical, soil, and environmental characteristics, a common situation is that surface processes occur at faster rates than subsurface processes. Thus, if a chemical is soil-incorporated or buried, it would degrade at a slower rate than if it were surface applied, and thus it would present greater potential for leaching to ground water.

For surface-applied chemicals that are non-persistent, the faster the chemical moves to the subsurface zones by soil-water movement, the greater the potential for reaching ground water because of the slower decay in the subsurface soil. This situation would be reversed if subsurface degradation processes occurred at rates faster than surface processes.

For persistent chemicals that demonstrate half-lives in soils on the order of 100 days or more, the relative timing of rainfall or irrigation events (i.e., with respect to placement and application) is less critical to chemical movement since the chemical resides in the soil for a much longer time period. However, the impact of surface and subsurface processes and of their rates is still a primary consideration.

Included within this category of transformation and degradation processes are those chemical or biological mechanisms which can transform a parent compound into various metabolites or daughter products which may be of equal or greater environmental concern due to mobility, persistence, or toxicity characteristics. Moreover, these characteristics of the metabolites may not be the same as those of the parent chemical. Experience with a number of agricultural chemicals [10,11,12,13] has shown the need to consider transformation products and their differing mobility and persistence when evaluating the environmental fate of a chemical.

Volatilization

Volatilization is generally defined as the loss of chemical in vapor form to the atmosphere from soil, plant, or water surfaces. Since our interest here is in the soil environment, volatilization from soil and plant surfaces is of primary concern. Since chemical applications in agriculture, silviculture, and land application of wastes (i.e., treatment or disposal) can be made to both bare soil and growing plants, volatilization from these types of surfaces can occur and can be important. In addition, volatilization of wastes buried in landfills and waste disposal sites can be a significant pathway for human exposure.

As a chemical fate process, volatilization can be a major loss mechanism resulting in reduced amounts of chemical in the soil that would be susceptible to leaching and surface runoff. The extent to which volatilization is important depends on chemical characteristics such as vapor pressure and Henry's Law constant. In addition, soil, environment, and management variables are important (as discussed in Chapter 2). For buried and soil-incorporated chemicals, volatilization controls the extent to which the chemical will exist in the vapor phase within the soil and thus will move through the soil by vapor diffusion. Consequently, volatilization includes aspects of both chemical fate and transport.

Plant Processes

Vegetation is an integral part of the terrestrial ecosystem that encompasses the soil environment. Chemicals applied to the land will enter the biological system of a plant and will undergo the full range of transport and fate mechanisms discussed above. In effect, the plant is a subsystem within the terrestrial ecosystem where chemicals enter, are transported, are sorbed onto organic carbon, are transformed and degraded, and, specific to biological systems, can accumulate in different portions of the plant. This bioaccumulation process is especially important since it has implications for direct exposure to animals and humans through the food chain.

Chemicals applied to the land surface, such as in agricultural or silvicultural operations or land application of wastes, can be intercepted by growing vegetation so that a portion of the applied chemical remains on the plant. On the plant leaf surfaces, fate processes such as photolysis, biodegradation, and volatilization can occur in addition to direct uptake and absorption into the body of the plant. During subsequent rainfall or irrigation events, the chemical can be dislodged and washed off the leaf surfaces to the soil [14].

From the soil, the plant roots allow uptake of the chemical into the plant where it can be transported or translocated among the roots, stems, and leaves. Within the body of the plant, the translocation process is affected by the partitioning of the chemical between the dissolved and sorbed phases [15]. Chemicals transported to the leaves can be exuded onto the leaf surface and can undergo the fate processes noted above. Accumulation of chemicals within the plant will vary for different portions such as the root, stems, leaves, fruit, etc.; subsequent potential exposure to animals and humans depends on the extent and location of the accumulation. When plants die, drop their leaves, or are incorporated into the soil, the remaining chemical may be returned to the soil where it is subject to all the soil fate and transport processes.

Thus, plant processes include a complex set of interacting mechanisms that control the chemical uptake, transformation, and fate within the plant system. For agricultural and silvicultural operations, and land application and treatment of wastes, plant processes can be a significant component of chemical fate and exposure. For landfills and buried waste disposal sites, these processes are likely to be negligible due to the lack of extensive vegetation and rooting systems.

Space and Time Variability of Soil Processes

A key aspect of the soil environment is its inherent variability in space and time. Spatial variability refers to changes in a measured value (e.g., soil property, solute concentration) at a specified depth throughout the area of interest (e.g., a field) as well as changes with depth at a specific location. Temporal variability is the change in a value with time for any specific location in the field. In effect, we are dealing with a fully three-dimensional dynamic system resulting from the indigenous spatial variation of soil properties and impacted by the temporal variation of natural processes and human intervention. These variations are comprised of both intrinsic factors, such as natural variations in soil characteristics, and extrinsic factors, such as water and chemical applications, tillage practices, etc. [16]. The processes described in the foregoing section demonstrate significant temporal and spatial changes as a result of this total variability, both intrinsic and extrinsic. These variations have important implications for collection of field measurements, interpretation of field data, and modeling of soil system processes.

Spatial variations in chemical behavior in soil systems are primarily a result of significant point-to-point changes in soil properties related to the structure, texture, composition, mineral content, organic content, etc. These properties have a direct impact on virtually all the parameters that characterize chemical transport and on many of the parameters that characterize chemical fate processes. In addition, these properties will influence environmental conditions - soil water content, temperature, pH - which in turn impact the chemical transport and fate processes.

For agricultural systems, additional variability is introduced by the specific agronomic practices that may impose a specific variation: for example, banded, surface-applied, or soil-incorporated chemical application; planting and tillage in rows; and furrow, drip or sprinkler irrigation [16]. Natural, spatial variations in meteorologic conditions, primarily rainfall, are subsequently superimposed as additional spatial variability.

For waste disposal systems, barriers to water and chemical movement, such as clay and synthetic liners, are deliberately constructed to contain the movement of the waste and prevent external contamination. Such barriers usually have relatively uniform properties that will likely be much different from the native soil (e.g., hydraulic conductivity). Chapter 11 provides a detailed discussion of the spatial variability of soil properties including current research and implications for field data collection and interpretation.

Temporal variations are also comprised of both intrinsic and extrinsic components. The major intrinsic factor is the natural variation of rainfall; although rainfall is also spatially variable (as noted above), its temporal variation is the primary driving force behind the dynamic nature of the soil environment and the resulting chemical movement. Other meteorologic inputs, such as temperature, solar radiation, wind, etc., are also highly time-variable and have significant impacts on chemical processes through their interdependent environmental conditions.

All of the soil chemical fate and transport processes discussed above demonstrate significant kinetic, or time-variable, behavior, often as a direct result of the time-variable inputs. The transport and sorption processes are thought to be relatively fast; they are initiated and occur during and immediately following water applications. The fate processes of transformation, degradation, and volatilization occur generally at a slower rate, but on a more continual basis. Thus, for agricultural chemical applications of nonpersistent chemicals, the time between chemical application and the first few rainfall events is a major determinant of the amount of chemical runoff and leaching. The longer this time period, the greater the opportunity for the chemical to dissipate through various pathways; thus, less of the chemical is available to runoff and leach. Chapter 2 further describes the kinetic nature of soil processes affecting chemical fate and migration.

Plant processes demonstrate a definite temporal pattern related to the natural crop growth cycle of planting, emergence, development, maturity, and harvest. These phases are in turn affected by the temporal variation of environmental conditions which jointly influence the chemical uptake, transformation, and fate processes within the plant.

The primary extrinsic factors that affect temporal variations in soil processes are related to the nature, timing, and frequency of management practices and operations. These activities include water and chemical applications, and disturbances to the soil as tillage, clear-cutting, or landfilling operations. The timing of water and chemical applications relative to evapotranspiration rates and the occurrence of natural rainfall events impact the movement of both water and chemicals. Soil disturbances cause changes to basic soil properties that affect runoff and infiltration which in turn affect associated chemical losses.

MODELING SOIL SYSTEM PROCESSES

Models, as used in this report, are simply representations or approximations of terrestrial environmental systems. They are not exact. They are simplifications of a real system since no model can realistically represent in detail the intricate workings and processes of real terrestrial systems described above. Models are basically "cartoons" of reality, as illustrated by the schematic representation in Figure 1-1; they attempt to represent the essential characteristics and behavior of a real system.

In general, models are used to analyze system behavior under both current (or past) conditions and anticipated (or future) conditions. Modeling requires some basic knowledge of the system being analyzed; however, it also promotes an improved understanding of the system through sensitivity analyses of system characteristics and observations of the resulting system response as predicted by the model and characterized by field data. This research-type function is especially important for soil leaching models to help expand our current knowledge of the complex terrestrial environment.

The most critical and cost-effective use of models is in the analysis of proposed or alternative future conditions. That is, the model is used as a management or decision making tool to help answer the "what if" type questions, such as "What if the application rate, timing, or method were changed?" or "What if management practices, remedial actions, or control measures are instituted?" Attempting to answer such questions through data collection programs would be enormously expensive and practically impossible in many situations.

Most mathematical models of environmental systems can be referred to as "linked-process models" since they include mathematical representations of individual processes and their interactions (or linkage) in order to approximate an entire environmental system. Figure 1-2 shows the hierarchy of model components. The model is comprised of linked or integrated processes which are mechanisms known to occur within the environmental system being modeled. An individual process is represented by an algorithm (or series of algorithms) which are in turn comprised of mathematical equations or expressions that combine to represent the process. Finally, the equations and expressions include variables and parameters. A variable is a time-dependent storage or flux in an algorithm; it can be an environmental variable such as an input meteorological condition (e.g., precipitation, air temperature, solar radiation) or an output variable which describes the behavior of the system (e.g.,

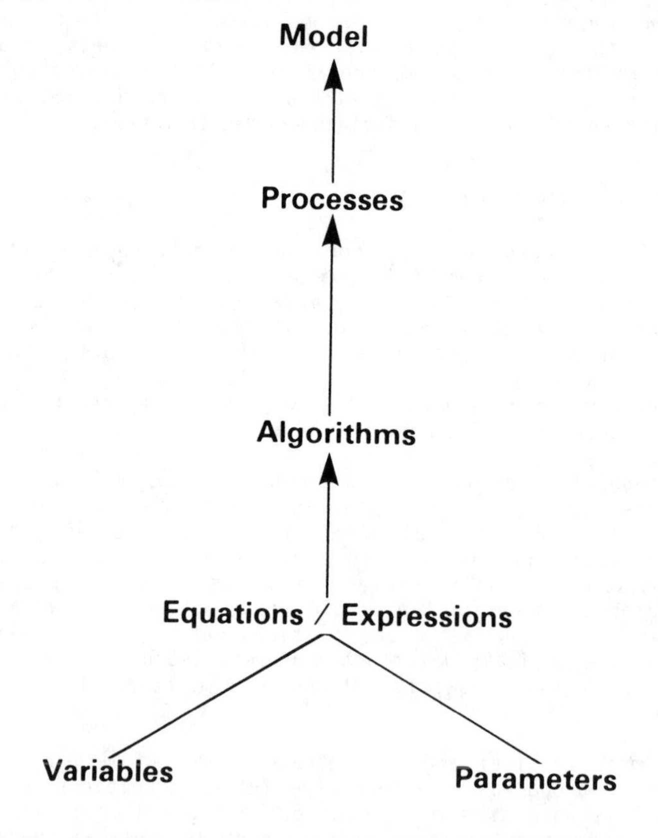

Figure 1-2. Hierarchy of model components (Reprinted with permission from A. S. Donigian annd J. D. Dean, 1985, "Nonpoint source pollution models for chemicals," in Environmental Exposure from Chemicals, W. B. Neely and G. E. Blau, Eds., Copyright, CRC Press, Inc., Boca Raton, FL).

flow, solute concentration, chemical storage). A parameter is a constant in an equation that describes a characteristic or property of the system (e.g., saturated hydraulic conductivity, sorption coefficient, hydrolysis rate) [14].

The American Society for Testing and Materials (ASTM) has prepared a Standard protocol for evaluating environmental chemical fate models, including the definition of selected modeling terms shown in Table 1-1 [17]. These terms are commonly used in the modeling literature although often with different meanings; the ASTM definitions in Table 1-1 may help to improve communications among modelers and with other professionals by providing more order to the technical language.

Model Classifications

Models to represent the fate of chemicals in the soil environment have proliferated during the past decade, primarily as a result of increased attention and concern for protecting groundwater supplies from contamination by hazardous wastes, pesticides, and other toxic chemicals. Unfortunately, a variety of often confusing terms has been used to categorize and differentiate the capabilities and characteristics of the many models available. Terms such as deterministic, empirical, stochastic, analytic, semianalytic, numerical, mechanistic, equilibrium, kinetic, etc. have been used sometimes differently by different authors in attempts to classify models. Moreover, the models have been developed by scientists and engineers with different backgrounds, perspectives, and for different purposes, (e.g., research versus regulatory or management), and often can be described by a number of the terms listed above. For example, a single model may take a deterministic approach to process representation, use a numerical procedure for solving the governing transport equation, assume equilibrium sorption, and describe biodegradation with first-order kinetics. Thus, any model which is an assemblage of interacting processes may include different approaches to best represent a specific process.

The primary differences between models are the level of detail at which the fundamental soil processes discussed in Chapter 2 are treated, and the conceptual completeness in terms of the number of processes of concern included in the model [18]. Inherent differences in models are also related to the manner in which temporal and spatial variations are or are not considered. Rao et al. [18] suggest that these primary differences in model characteristics are determined in part by the following factors:

a. Current state of understanding of the system and its components (subsystems) to be modeled.

b. The modeler's conceptualization of the system processes.

c. The modeling approach and error bounds allowable in approximations required to solve the problem.

Table 1-1. Definition of Terms Used in ASTM Standard Practice for Evaluating Environmental Fate Models of Chemicals.

Algorithm -	the numerical technique embodied in the computer code.
Calibration -	a test of a model with known input and output information that is used to adjust or estimate factors for which data are not available.
Compartmentalization -	division of the environment into discrete locations in time or space.
Computer Code (computer program) -	the assembly of numerical techniques, bookkeeping, and control language that represents the model from acceptance of input data and instructions to delivery of output.
Model -	an assembly of concepts in the form of a mathematical equation that portrays understanding of a natural phenomenon.
Sensitivity -	the degree to which the model result is affected by changes in a selected input parameter.
Validation -	comparison of model results with numerical data independently derived from experiments or observations of the environment.
Verification -	examination of the numerical technique in the computer code to ascertain that it truly represents the conceptual model and that there are no inherent numerical problems with obtaining a solution.

Source: ASTM [17].

d. The spatial and temporal scales of intended model applications.

Although the terms used in classifying models leave much to be desired, some discussion of the major model types is needed to provide a framework for relating the models described in this report to the universe of models and modeling approaches. One such classification proposed by Addiscott and Wagenet [19] is shown in Table 1-2 and includes references to models represented by each category. The two major groups in this classification are deterministic and

Table 1-2. A Classification of Leaching Models.

I. Deterministic models

A. Mechanistic (usually based on rate parameters)

1. Analytical* [e.g., 20,21].

2. Numerical* [e.g., 22,23].

B. Functional (usually based on capacity parameters)

1. Partially analytical [e.g., 24,25,26].

2. Layer and other simple [e.g., 27,28,29,30].

II. Stochastic Models

A. Mechanistic [e.g., 31,32].

B. Non-mechanistic (transfer function) [33,34].

*Refers to the solution of the flow equations.

Source: Addiscott and Wagenet [19].

stochastic with the differences described by Addiscott and Wagenet as follows:

> A key distinction in comparisons between models is that between deterministic models, which presume that a system or process operates such that the occurrence of a given set of events leads to a uniquely-definable outcome, and stochastic models, which presuppose the outcome to be uncertain and are structured to account for this uncertainty.

The vast majority of soil chemical fate models are deterministic since they attempt to represent the primary physical, chemical, and biological processes that determine chemical fate and movement in soil systems. In fact, most of the so-called stochastic models are not completely stochastic in that they often utilize a deterministic representation of soil processes and derive their stochastic nature from their representation of inputs and/or spatial variation of soil characteristics and resulting chemical movement. While the deterministic approach results in a specific value of a soil variable (e.g., solute concentration) at any point in a field, the stochastic approach provides the probability (within a desired level of confidence) of a specific value occurring at any point [16].

Development of stochastic models is a relatively recent endeavor that has occurred as a result of the growing awareness of the importance of intrinsic and extrinsic variability of the soil environment. Consequently, stochastic models are still primarily research tools, although they may hold promise for management purposes in the future [19].

Attempts to validate terrestrial leaching models must address the issue of spatial variability when comparing model predictions with limited field observations. If sufficient field data are obtained to derive the probability distribution of pesticide concentrations, the results of the stochastic model can be compared directly. For a deterministic model, the traditional approach has been to vary the input data within its expected range of variability (or uncertainty) and determine whether the model results fall within the bounds of field measured values. Given the lack of comprehensive field data sets that adequately characterize spatial variability, the absence of widely accepted statistical methods for judging the goodness-of-fit models to measured data, and our inability to directly measure water and solute fluxes which are more logical variables for model validation, it has been suggested that complete validation of any simulation model may not be possible [16]. Such complete validation may not be necessary for certain model cases if model limitations are adequately recognized.

The second level of classification in Table 1-2 is mechanistic vs. functional for deterministic models and mechanistic vs. non-mechanistic for stochastic models. The distinction between mechanistic and functional is really a matter of the degree of process representation; mechanistic models incorporate the most fundamental understanding of soil processes, such as the use of Richard's equation (see Chapter 6) to define soil water movement, whereas functional models use more simplified representations, such as field capacity concepts for soil moisture retention. Functional models are largely management-oriented while mechanistic models are primarily research-oriented.

The stochastic-mechanistic models as noted above combine elements of both approaches. The stochastic non-mechanistic models are purely statistical and must rely on some degree of calibration for evaluation of selected model parameters. Although still developmental, this approach may hold promise in the future for management tools with consideration of variability and uncertainty [19].

Another distinction between model types listed in Table 1-2 is analytical versus numerical deterministic models. Although Table 1-2 lists this under mechanistic-deterministic models, it really applies to the entire category of deterministic models. Analytical models are based on a direct solution to the governing convective-dispersive solute transport equation (see Chapter 6) under defined initial and boundary conditions and on other assumptions. Certain conditions and assumptions are required in order to obtain a direct solution to the equation. The most critical of these assumptions for analytical soil leaching models are as follows:

1. Constant, uniform soil-water content and soil-water flux throughout the soil profile (steady water flow).

2. Uniform, homogeneous soil properties.

3. Constant, pulse, or exponentially decaying chemical boundary conditions at the soil surface.

In spite of these simplifying restrictions, analytical models have been used extensively in a variety of environmental applications. Van Genuchten and Alves [35] have cited a number of example applications of analytical models in soil science, chemical and environmental engineering, and water resources and have provided a useful compendium of solutions under various combinations of initial and boundary conditions.

Numerical models are designed to provide the flexibility to analyze dynamically changing and heterogeneous soil conditions that are often found in field applications. They utilize numerical solution techniques (e.g., finite difference, finite element) to solve the governing solute transport equation. Thus, numerical models are designed to circumvent the restrictive assumptions of the analytical models.

Review and Comparison of Selected Models

The primary models considered in this report are the following three models developed or sponsored by various EPA offices: PESTAN, SESOIL, and PRZM. Each of these models is described below and is followed by a general comparison with other available models.

PESTAN

PESTAN is an interactive program developed by the EPA R. S. Kerr Environmental Research Laboratory in Ada, Oklahoma, for estimating the movement of organic chemicals through soil to ground water. The theory on which PESTAN is based is described by Enfield et al. [36], who discuss three simple models derived as primarily analytical solutions to the convective-dispersive solute transport equation. In addition to the vertical (i.e., one-dimensional) convective movement of a chemical, these three models include the following specific combinations of processes:

1. Linear sorption and first-order decay (transformation), no dispersion

2. Dispersion and linear sorption, no decay (transformation)

3. Non-linear (Freundlich) sorption and first-order decay (transformation) no dispersion

Models 1 and 2 are complete analytical solutions similar to those tabulated by van Genuchten and Alves [35], while model 3 uses the method of characteristics to obtain a numerical solution to the governing equation due to the use of a non-linear isotherm. Note that none of the three models account for vapor-phase partitioning and transport. Thus, PESTAN's use is limited to nonvolatile chemicals.

The PESTAN program is the computer implementation of model 2 described by Enfield et al. [36], with the additional capability to consider first-order decay processes within an interactive framework that simplifies model use and application. The primary output of the model is the chemical concentration distribution within the soil profile for any user-specified time after application. The program is easy to use, requires evaluation of only 10 input parameters, and can be run interactively from a computer terminal without the need for a user manual. PESTAN has been used extensively by the EPA Office of Pesticide Programs (OPP) for initial screening assessments to evaluate the potential for ground-water contamination of pesticides submitted for registration and for currently registered compounds [37]. It has also been tested against field data by the authors [36] and by other investigators [11,38,39].

In spite of its simplicity, PESTAN can be a useful tool for preliminary assessments of the type performed by OPP as long as the user is fully aware of its assumptions and limitations. The primary limitations of PESTAN are directly related to its analytical nature which requires the assumption of one-dimensional steady water flow in a homogeneous soil profile with constant hydraulic, sorption, and decay parameters. Thus, the temporal variability of soil processes, especially leaching, due to the relative timing and intensity of water and chemical application rates, and the spatial variability (primarily vertical heterogeneities such as layering) of soil characteristics are largely ignored. The combined result of these assumptions can lead to underestimation of subsurface chemical concentrations and the resulting contaminant input to ground water [11,37,38]. PESTAN may be most appropriate for the intermediate unsaturated zone between the bottom of the root zone and ground water because the steady flow assumption is most applicable in this region. However, it should be noted that decay of the contaminant within the vadose zone is still not accounted for in Model 2.

PESTAN users should perform sensitivity analysis on key parameters, such as water recharge rate, decay rate, and sorption coefficient, in order to assess the potential impact of significant variations in these parameters under field conditions. The water recharge rate is a key parameter determining the rate of chemical movement to ground water. Simple water balance (i.e., annual precipitation and irrigation minus evaporation) calculations may significantly underestimate this parameter [11]. Also, users should be careful in evaluating the input decay rate required by PESTAN; it is defined on the basis of the sorbed chemical concentration as opposed to the total concentration. Thus, total soil persistence or attenuation rates must be appropriately adjusted prior to input [37].

SESOIL

The SEasonal SOIL Compartment (SESOIL) model was developed for EPA Office of Toxic Substances (OTS) in response to the need for a model for long-term simulation of chemical fate in the soil environment. The current version of SESOIL is designed for "... environmental process simulations that can describe simultaneously water transport, sediment transport, and pollutant fate (transport/transformation)...in an unsaturated soil compartment" [40]. A unique feature of SESOIL is that the hydrology is based on a statistical representation of the water balance components over a "season" rather than accounting for changes in soil moisture through time.

According to Bonazountas and Wagner [41], the intended uses or applications of SESOIL include "...long-term leaching studies from waste disposal sites, acid rain, pesticide and sediment transport on watersheds, contaminant exposure, pre-calibration runs for other simulation models, hydrologic cycles and water balances of soil compartments, overall chemical fate assessments, exposure, risk and other studies."

SESOIL is a seasonal model in that it calculates the pollutant distribution in the soil column and on the watershed at the end of a "season" (e.g., year or month). The model is compartmental in that the soil column can be subdivided into at least four layers or compartments with each having uniform properties. Although each layer (compartment) is assumed to have uniform properties, the depth-varying nature of soil properties can be approximated by assigning an average or weighted value to each layer.

SESOIL is structured for the integrated simulation of three cycles: the hydrologic cycle, the sediment cycle, and the pollutant cycle. Each cycle encompasses a multitude of processes as shown in Table 1-3; some of these processes are not operational in the latest version of SESOIL.

The hydrologic cycle in SESOIL is based on a statistical dynamic formulation of the vertical water budget at the land-atmosphere interface [42]. Physically based dynamic and conservation equations express the infiltration, exfiltration, transpiration, percolation to ground water, and the capillary rise from the water table during and between storms in terms of independent variables of precipitation, potential evapotranspiration, soil properties, and water table elevations. Uncertainty of the hydrologic cycle simulation is introduced into these equations via the probability density functions of the independent climatic variables and produces derived probability distributions of the dependent water balance elements: surface runoff, evapotranspiration, and ground-

Table 1-3. Simulation Capabilities of SESOIL Cycles.

Hydrology	Sediment	Pollutant
precipitation	resuspension by wind	advection
infiltration/ exfiltration	washload	diffusion
soil moisture		volatilization
evapotranspiration		sorption
capillary rise		degradation
ground-water runoff		hydrolysis
surface runoff		oxidation
snow pack/melt*		cation exchange
interception*		complexation chemistry
		photolysis*
		nutrient cycles*
		biological uptake*

*Not operational in current version of SESOIL.

water runoff. The mean value of these quantities give a long-term (season) average water balance.

The sediment cycle includes simulation of both sediment washload due to precipitation runoff and sediment (dust) resuspension due to wind. Options for annual and monthly simulation of sediment washload are included. The annual option is based on the Universal Soil Loss Equation [43], while the monthly option uses modifications of procedures developed by Foster et al. [44] for the CREAMS model [45]. Resuspension losses from the surface compartment due to physical removal of particles and associated pollutants as a function of particle characteristics (chemical composition, diameter, etc.) and weather/soil conditions (e.g., wind speed, soil moisture) are also considered.

The pollutant cycle allows for simulation of the 12 chemical processes noted in Table 1-3; some alternate simulation options are available for selected processes. Diffusion is described by Fick's first law to estimate upward solute flow in the soil column to the air as volatilization. Volatilization considers pollutants buried under a layer of clean soil, assumes Fick's first law, and assumes vapor phase diffusion is rate controlling. Adsorption/desorption is represented by the Freundlich isotherm, and degradation is assumed to follow first-order kinetics. Neutral and acid or base catalyzed hydrolysis is also described as first-order decay. Cation exchange is assumed to be an irreversible, instantaneous process which immobilizes pollutant up to the soil cation exchange capacity. Consideration is also given to complexation of metal ion in solution with organic ligand(s).

SESOIL has undergone testing by a variety of organizations in addition to that done by its developers. These efforts have included sensitivity analysis, comparison with other models, and some limited comparisons with field data. Arthur D. Little, Inc., the developers of SESOIL, compared model predictions with field data for selected metals at a site in Kansas, and for two organic compounds (i.e., naphthalene and anthracene) at a site in Montana [46]. Sensitivity analyses on the volatilization and adsorption routines were performed by Wagner et al. [47], for six pollutants in three climates with four different soil types. Hetrick [48] compared SESOIL hydrologic predictions with some calibration to analogous results from a more detailed hydrologic watershed model, AGTEHM [49], and to observations from a watershed in Tennessee and South Carolina. He concluded that SESOIL hydrologic predictions were in good agreement with both AGTEHM predictions and observed data but that the month-to-month deviations indicated that SESOIL should not be recommended for short-term predictions [48].

SESOIL was incorporated as the soil/land component of the screening level multimedia model, TOX-SCREEN [50], developed by Oak Ridge National Laboratories for the EPA Office of Toxic Substances. In a separate model evaluation study, Bicknell et al. [51] attempted to test TOX-SCREEN with multimedia data on benzo(a)pyrene from Southeastern Ohio. However, no definitive conclusions on the performance of SESOIL could be made because model predictions could not be directly compared with measured soil concentrations.

Battelle Pacific Northwest Laboratories recently completed a model comparison study that included SESOIL along with a number of flow, transport, and geochemical models for soil and ground-water systems [52]. SESOIL has also been compared in a laboratory column study with PESTAN and PRZM [39]. A landfill test problem was designed for comparing flow predictions for one-dimensional unsaturated zone models. This latter study concluded that SESOIL was not useful for analyzing sites with a high degree of vertical variation in soil properties (i.e., layering) because SESOIL characterizes the soil column as homogeneous and isotropic for water balance calculations.

The most recent and comprehensive evaluation of SESOIL was performed by Watson and Brown [53]. The primary conclusions derived from that study are as follows:

1. The basic framework for a useful, screening-level, chemical migration and fate model exists in SESOIL. However, there are several errors in the pollutant cycle (i.e., relative to neutral hydrolysis and chemical leaching) that require correction, and several modifications should be considered to improve the representation of chemical leaching and washoff from the land surface.

2. SESOIL can be applied to generic environmental conditions if model limitations are observed and judgment is used in the estimation of model parameters. As with most models, SESOIL cannot be applied on a site-specific basis with only "limited calibration."

3. SESOIL cannot be applied to sites exhibiting large vertical variations in soil properties. Although up to four soil layers can be used for the pollutant cycles, a single homogeneous soil column is used in the hydrologic cycle.

4. Consideration should be given to modifying the algorithm used to determine when chemical will begin to leach to ground water. This algorithm is based only on the estimated travel time for water; the sorptive properties of the chemical are not considered.

5. Although the SESOIL documentation report/user's manual provides an extensive review of the literature in many areas, it is not explicit in terms of the specific processes that are considered, the theory that is incorporated into the model, and parameter estimation procedures.

6. There are limitations associated with using SESOIL as a watershed model, with applying SESOIL to certain disposal facilities (e.g., landfills), and with applications of SESOIL to certain release scenarios (e.g., spills, buried drums and tank leaks).

7. Although the newest version of SESOIL (dated May, 1984) is an improvement over the version tested by Watson and Brown [53], the improvements did not significantly change the testing conclusions.

SESOIL continues to be used and supported by the Office of Toxic Substances (OTS). Code corrections and enhancements are currently being performed by Oak Ridge National Laboratory as part of its continuing work on the TOXSCREEN model for OTS [54]. New users of SESOIL should review the testing report by Watson and Brown [53] to get further details on the model limitations noted above. A new, non-EPA version of SESOIL is currently under development at the Center for Environmental Management of the University of Athens in Greece. This new version, expected to be released in late 1985, will be based on a multiple layering scheme for representing the soil column, will include entirely new procedures for the water balance calculations, and will improve the sediment washload and evapotranspiration routines [55].

PRZM

The Pesticide Root Zone Model (PRZM), developed at the U.S. EPA Environmental Research Laboratory in Athens, Georgia [56], is a dynamic, compartmental model for use in simulating chemical movement within and below the plant root zone. Time varying transport including advection and dispersion are represented in the program.

PRZM has two major components: hydrology and chemical transport. The hydrology component for calculating runoff and erosion is based on the Soil Conservation Service curve number technique and the Universal Soil Loss Equation [43]. Evapotranspiration is estimated from pan evaporation or by an empirical formula if input pan data are unavailable. Evapotranspiration is comprised of evaporation from plant interception, evaporation from soil, and transpiration from the crop. Water movement is simulated by the use of an empirical model based on soil-water capacity terms including field capacity, wilting point, and saturation. To produce soil water and solid phase concentrations, the chemical transport component calculates pesticide uptake by plants, surface runoff, erosion, decay/transformation, vertical movement (leaching), foliar loss, dispersion, and retardation.

PRZM allows the user to perform dynamic simulations of potentially toxic organic chemicals, particularly pesticides, that are applied to the soil or to plant foliage. Dynamic simulations allow the consideration of pulse loads, the prediction of peak events, and the estimation of time-varying mass emission or concentration profiles; thus dynamic simulations overcome limitations of the more commonly used steady-state models.

To apply PRZM, the soil profile is divided into a number of soil layers or compartments. For each compartment, the model solves the solute transport equation (see Chapter 2) including advection, dispersion, adsorption, degradation, and plant uptake of the chemical. For the surface compartment, additional terms are included in the equation to allow for chemical losses associated with surface runoff and erosion. A numerical finite-difference solution procedure is used to solve the system of equations under appropriate boundary conditions.

Currently in PRZM, model soil parameters such as the sorption coefficient and degradation rate coefficient can be specified separately for each user-defined soil zone, e.g., surface zone, root zone, and below root zone. Each zone is then divided into uniform layers or compartments. As many soil layers and compartments can be utilized as are necessary to accurately represent the characteristics of the soil profile.

PRZM uses the Soil Conservation Service runoff curve number technique to distribute the daily rainfall into runoff and infiltration components. Because of this, the timestep for the model is one day. Infiltrating water is assumed to cascade downward to successively deeper layers as the soil water content of each compartment reaches and exceeds field capacity. Field capacity is usually reported as the soil-water content that field soils attain after all excess water is drained by gravity. For loose sandy soils, the water in excess of field capacity is assumed to drain within the one day time-step of the models. For tight clay soils and profiles with restrictive, low permeability layers, a drainage rate parameter can be specified by the user to allow excess drainage to occur over a longer time period (e.g., weeks or months). Using these procedures, the pore-water velocity is estimated and is not directly based on soil hydraulic conductivity; field capacity and wilting point values are used operationally to determine percolation and soil-water content in a compartment based on the infiltrating water. Wilting point is defined as the soil-water content below which plants are unable to extract water.

Evapotranspiration is calculated as a function of soil-water conditions and total potential evapotranspiration. It is either input by the user (and usually estimated from local pan evaporation data) or calculated from air temperature and the number of daylight hours.

The daily evapotranspiration demand is divided among evaporation from canopy, soil evaporation, and crop transpiration. Total demand is first estimated and then extracted sequentially from crop canopy storage and from each layer until wilting point is reached in each layer or until total demand is met. Evaporation occurs down to a user-specified depth. The remaining demand (crop transpiration) is met from the layers between this depth and the active rooting depth. The root zone growth function is activated at crop emergence and increases step-wise until maximum rooting depth is achieved at crop maturity.

Snowmelt calculations are included in PRZM based on the "degree-day" approach utilizing average daily air temperature. Runoff is then calculated using both rainfall and snowmelt as inputs to the curve number procedure. Based on the calculated daily runoff, PRZM estimates daily storm event erosion using the Modified Universal Soil Loss Equation (MUSLE) developed by Williams [57]. The peak storm runoff rate required by MUSLE is estimated by assuming a trapezoidal hydrograph and a mean storm duration input by the user.

Degradation of the chemical is represented as a single first-order process with the rate coefficient specified by the user for each defined soil zone. Thus, representation includes all significant biochemical transformation and decay processes that reduce the amount of chemical in the soil. Different rates are commonly used for the surface zone, root zone, and below root zone (partially) to account for the different transformation/degradation processes occurring in the various portions of the profile. Plant uptake is represented as a separate process and is calculated as a function of the amount of transpiration in each compartment and as an uptake efficiency factor.

Like SESOIL and PESTAN, PRZM does not account for vapor-phase partitioning and transport (via vapor diffusion). Thus, all three models cannot be used to evaluate data for volatile chemicals (e.g., EDB, DBCP, TCP, etc.).

PRZM has undergone a moderate degree of performance testing in comparison with field data at selected sites across the country and has been used in exposure assessments and other applications. Because of its agricultural emphasis and the more general availability of field pesticide data, the primary testing studies listed below are for pesticide leaching.

a. Aldicarb applied to citrus in Florida [11]

b. Aldicarb applied to potatoes in New York [58] and Wisconsin [59]

c. Metalaxyl applied to tobacco in Florida and Maryland [60]

d. Atrazine and chloride applied to corn in Georgia [58]

The results of these testing efforts by both EPA (i.e., the model developers) and other investigators have been consistently positive when comparing field average values to PRZM predictions of soil profile concentrations and mass flux to ground water. The spatial variability problems noted earlier preclude effective comparison between PRZM results and concentrations from individual points within the field. Considering the relatively simplistic nature of PRZM relative to the dynamic soil environment, the testing results to date demonstrate that PRZM is effectively representing the primary processes controlling pesticide movement to ground water [60].

The PRZM model has been requested by and distributed to over 150 users nationwide and is currently being applied in a variety of locations. The EPA Environmental Research Laboratory in Athens, Georgia, is heading a cooperative research project with the U.S. Geological Survey to develop field data for further testing and refinement of PRZM and for investigating the spatial variability of pesticide leaching in field soils [60; see Chapter 4 for a description of this project].

Since its initial release in "draft" form in 1982, PRZM has been used in a wide range of studies. Both the EPA Office of Pesticide Programs (OPP) and various chemical manufacturers have used PRZM to assess the leaching potential of new and currently registered compounds [61]. Dean et al. [62], used PRZM to develop a methodology allowing expeditious use by OPP to screen and assess pesticides for potential contamination of ground water on a national scale; running PRZM for 25 years at selected sites across the country, probability distributions of pesticide leaching below the root zone were developed as a function of pesticide characteristics and cropping practices.

PRZM has also been used as a framework for a terrestrial ecosystem model for pesticide exposure to wildfowl [63], has been linked to ground water, surface water, and air models for a multimedia study of the effects of conservation tillage on pesticide concentrations [64], and has provided the key element of a PCB spill exposure assessment methodology [65].

In conjunction with detailed multidimensional ground-water models, PRZM was used as part of an exposure assessment for aldicarb carb application to citrus in Florida [66] and as part of an analysis of the effects of facility design and locational factors on leachate migration from hazardous waste sites [67].

Model Comparisons

In order to provide background on the general types of models available, Table 1-4 summarizes and compares the primary capabilities and characteristics of PESTAN, SESOIL, and PRZM along with a few additional models of flow and contaminant movement in the unsaturated zone.

CREAMS2 [68] is a recent modification and improvement of the USDA CREAMS model [45] originally developed for analysis of agricultural runoff. CREAMS2 includes capabilities to model chemical movement through the crop root zone using a layered, compartmental approach [69].

The HELP (Hydrologic Evaluation of Landfill Performance) model was developed for use by permit writers and landfill designers to assess water movement across, into, through, and out of landfills with rapidity [70]. It was recently modified to consider contaminant movement by volatilization and percolation of the leachate through the landfill [71].

The remaining models listed in Table 1-4 are representative of a large group of detailed numerical models that have been developed for variably saturated conditions (i.e., unsaturated and saturated) and for multi-dimensional contaminant migration problems. This allows a more detailed representation of the dynamically changing interface between the unsaturated zone and ground water and requires more detailed information on the spatial variation of soil properties for the specific system being analyzed. These models generally include linear or Freundlich sorption isotherms, lumped first-order degradation, and simplified surface conditions for rainfall and evapotranspiration; runoff, erosion, and plant processes are usually ignored.

The last five models listed in Table 1-4 were selected for further study in a recent review of 55 flow and transport models for the unsaturated zone by Oster [72]. Only these five models were selected due to their consideration of both flow and transport, availability of code, documentation, and demonstrated application. Battelle, Pacific Northwest Laboratories recently completed an extensive evaluation of hydrogeochemical models for solute migration in both the unsaturated and saturated zones [52,73]; the five models noted above, plus SESOIL and CREAMS, were included in this evaluation. Interested readers are referred to

Table 1-4. Comparison of Selected Soil Leaching Models.

	TRANSPORT				TRANS. DEGRAD.			PLANT CHEMICAL PROCESSES								
	RUNOFF	EROSION	LEACHING	SORPTION	LUMPED	PROCESS-ORIENTED	VOLATILIZATION	UPTAKE	DEGRADATION	INTERCEPTION AND WASHOFF	VOLATILIZATION	TRANSLOCATION AND BIOCONCENTRATION	SPATIAL	NO. OF VERTICAL LAYERS	TIME STEP	ANALYTICAL VS. NUMERICAL
PRIMARY MODELS																
PESTAN			*	*	*								1	1	-	A
SESOIL	*	*	*	*	*	*	*						1	4	monthly	N
PRZM	*	*	*	*	*		a	*	*	*		a	1	user defined	daily	N
OTHER AVAILABLE MODELS																
CREAMS 2	*	*	*	*	*				*	*			1	7	daily	N
HELP/HELP-WQ	*		*	*	*		*						1	9		
SEGOL			*	*	*								2,3	-	variable	N
SUMATRA-1			*	*	*								1	-	variable	N
TARGET			*	*	*								1,2,3	-	variable	N
FEMWATER/FEMWASTE			*	*	*								2	-	variable	N
TRUST/MLTRAN			*	*	*								1,2	-	variable	N

a - These processes were included in an earlier experimental version of PRZM [63].

these two reviews for more detailed information on the extent of available soil leaching models.

Criteria for Model Selection and Use

As noted above, many models in addition to the 11 models listed in Table 1-4 are currently available for simulating organic pollutant fate and transport in the soil environment. The level of detail and the conceptual completeness of the available models vary considerably. As evident from the earlier sections of this chapter, these models are generally quite complex and include many processes and parameters. In selecting a model, a prospective user needs first to thoroughly understand the simplifications and assumptions made in the development of the model. Without sufficient documentation of the model, reports of adequate testing by independent workers, and a certain amount of hands-on experience, most prospective users and even many modelers would have considerable difficulty in appreciating all the intricacies and idiosyncracies of a model or of the computer code itself. This makes it difficult for a user to select among the models for his application. The following discussion, based primarily on a paper by Rao et al. [18], attempts to provide general guidelines for selection and use of simulation models. It is not intended as a set of steps to follow, but it does provide some of the factors to be considered by the user in model selection.

Rao et al. [18], suggest that the following evaluation factors be employed by a prospective user:

1. The intended use and the spatial/temporal scales at which the model simulations are to be used.

2. The availability of computational facilities (i.e., computer hardware, computer programmers, etc.) required to implement the computer code.

3. The availability and the reliability of the required model input data.

4. The confidence region(s) associated with the model output.

Two model attributes -- the type and scale of model output -- determine whether a particular model may be used without any modifications for a specific application. Each model and its computer code have a characteristic scale at which the model may be used. For example, a model designed for management purposes (say at a river basin scale) will most likely not be useful for detailed descriptions of pollutant fate at a small experimental field site. A model designed for research applications can be used for simulation at larger temporal and spatial scales. However, the prohibitive costs, the paucity of data for a large number of parameters in the model, and the spatial/temporal variability constraints would make such an application impractical if not impossible.

In many cases, the values of several model parameters are "adjusted" or "calibrated" when independent estimates are unavailable. It must be recognized that the number of parameters that are "adjusted" increases, and so may the uncertainty in their values. Whether the calibrated values of model parameters are appropriate for another site, season, or scale needs to be evaluated carefully.

Uncertainties in the model input data provided by the user may propagate through the various computations in a model and will be reflected in the model outputs. Such uncertainties, however, may not always have amplified effects (i.e., additive or multiplicative) on the model output data and may cancel each other out. With increasing complexity of the model, the user's ability to identify and rectify such problems, when present, is reduced.

Spatial and temporal variations in various processes and environmental factors also introduce uncertainties in measured data to be used for comparison with model output. Since well-established statistical (or other) procedures are presently unavailable, comparison of measured data with model predictions remains somewhat qualitative, and judgment of the goodness-of-fit is, in most cases, arbitrary. This, in our view, is probably the most vexing problem in model development, testing, and use.

All three models (PESTAN, SESOIL, and PRZM) selected for detailed discussions here were either developed by or sponsored by the USEPA. These models in general are well-documented and have been tested by a number of independent users. These models have been and will continue to be used by the EPA, state regulatory agencies, consulting firms, and industry. The discussion of the various processes and how they may be incorporated in a model as well as issues pertaining to parameter estimation and the design of field studies are intended to be "generic" and generally applicable to other simulation models.

REFERENCES

1. Rao, P. S. C., and R. E. Jessup. "Development and Verification of Simulation Models for Describing Pesticide Dynamics in Soils," Ecol. Modeling 16:67-75 (1982).

2. Wagenet, R. J., and P. S. C. Rao. "Basic Concepts of Modeling Pesticide Fate in the Crop Root Zone," Weed Sci. 33 (Suppl. 2): 25-32 (1984).

3. Baker, J. L. "Agricultural Areas as Nonpoint Sources of Pollution," in Environmental Impact of Nonpoint Source Pollution, M. R. Overcash and J. M. Davidson, Eds. (Ann Arbor, MI: Ann Arbor Science, 1980).

4. Wauchope, R. D. "The Pesticide Content of Surface Water Draining from Agricultural Fields - A Review," J. Environ. Qual. 7:459-471 (1978).

5. Wauchope, R. D., and R. A. Leonard. "Maximum Pesticide Concentrations in Agricultural Runoff: A Semiempirical Prediction Formula," J. Environ. Qual 9:665-672 (1980).

6. Baker, D. B. "Studies of Sediment, Nutrient and Pesticide Loading in Selected Lake Erie and Lake Ontario Tributaries," EPA Grant No. R005708-01 (1983).

7. Johnson, H. P. and J. L. Baker. "Field-to-Stream Transport of Agricultural Chemicals and Sediment in an Iowa Watershed: Part I. Data Base for Model Testing (1976-1978)," EPA-600/3-82-032 (Athens, GA: U.S. EPA, 1982).

8. Smith, C. N., R. A. Leonard, G. W. Langdale, and G. W. Bailey. "Transport of Agricultural Chemicals from Small Upland Piedmont Watersheds," EPA-600/3-78-056, IAG No. IAG-D6-0381 (Athens, GA: U.S. EPA and Watkinsville, GA: USDA, 1978).

9. Ellis, B. G., A. E. Erickson, A. R. Wolcott, M. Zabik, and R. Leavitt. "Pesticide Runoff Losses from Small Watersheds in Great Lakes Basin," EPA-600/3-77-112 (Athens, GA: U.S. EPA, 1977).

10. Hornsby, A. G., P. S. C. Rao, W. B. Wheeler, P. Nkedi-Kizza, and R. L. Jones. "Fate of Aldicarb in Florida Citrus Soils: 1. Field Laboratory Studies," Proceedings of the NWWA/U.S. EPA Conference on Characterization and Monitoring of the Vadose (Unsaturated) Zone, Las Vegas, NV (1983).

11. Jones, R. L., P. S. C. Rao, and A. G. Hornsby. "Fate of Aldicarb in Florida Citrus Soil: 2. Model Evaluation," Proceedings of the NWWA/U.S. EPA Conference on Characterization and Monitoring of the Vadose (Unsaturated) Zone, Las Vegas, NV (1983).

12. Dean, J. D., D. W. Meier, B. R. Bicknell, and A. S. Donigian, Jr. "Simulation of DDT Transport and Fate in the Arroyo Colorado Watershed, Texas" (Athens, GA: U.S. EPA, 1984).

13. Bilkert, J. N., and P. S. C. Rao. "Sorption and Leaching of Three Non-Fumigant Nematicides in Soils," J. Environ. Sci. Health B20:1-26 (1985).

14. Donigian, A. S. Jr., and J. D. Dean. "Nonpoint Source Pollution Models for Chemicals," in Environmental Exposure from Chemicals, Vol. 2, W. B. Neely and G. E. Blau, Eds. (Boca Raton, FL: CRC Press, Inc., 1985).

15. Briggs, G. G., R. H. Bromilow, and A. A. Evans. "Relationships Between Lipophilicity and Root Uptake and Translocation of Non-Ionized Chemicals by Barley," Pest. Sci. 13:495-504 (1982).

16. Rao, P. C. S., and R. J. Wagenet. "Spatial Variation of Pesticides in Field Soils: Methods for Data Analysis and Consequences," Weed Sci. 33: (In Press).

17. ASTM. Standard Practice for Evaluating Environmental Fate Models of Chemicals. (American Society for Testing and Materials, 1984).

18. Rao, P. S. C., R. E. Jessup, and A. G. Hornsby. "Simulation of Nitrogen in Agroecosystems: Criteria for Model Selection and Use," Plant and Soil 67:35-43 (1982).

19. Addiscott, T. M., and R. J. Wagenet. "Concepts of Solute Leaching in Soils: A Review of Modeling Approaches," J. Soil Sci. 36:411-424 (1985).

20. Nielsen, D. R., and J. W. Biggar. "Miscible Displacement. III. Theoretical Considerations," Soil Sci. Soc. Amer. Proc. 26:216-221 (1962).

21. van Genuchten, M. Th., and P. J. Wierenga. "Mass Transfer Studies in Porous Sorbing Media. I. Analytical Solutions," Soil Sci. Soc. Amer. J. 40:473-480 (1976).

22. Childs, S. W., and R. J. Hanks. "Model for Soil Salinity Effects on Crop Growth," Soil Sci. Soc. Amer. Proc. 39:617-622 (1975).

23. Robbins, C. W., R. J. Wagenet, and J. J. Jurinak. "A Combined Salt Transport-Chemical Equilibrium Model for Calcareous and Gypsiferous Soils," Soil Sci. Soc. of Amer. J. 44:1191-1194 (1980).

24. De Smedt, F., and P. J. Wierenga. "Approximate Analytical Solution for Solute Flow During Infiltration and Redistribution," Soil Sci. Soc. Amer. J. 42:407-412 (1978).

25. Rose, C. W., F. W. Chichester, J. R. Williams, and J. T. Ritchie. "A Contribution to Simplified Models of Field Solute Transport," J. Environ. Qual. 11:146-150 (1982).

26. Rose, C. W., F. W. Chichester, J. R. Williams, and J. T. Ritchie. "Application of an Approximate Analytic Method of Computing Solute Profiles with Dispersion in Soils," J. Environ. Qual. 11:151-155 (1982).

27. Bresler, E. "A Model for Tracing Salt Distribution in the Soil Profile and Estimating the Efficient Combination of Water Quality and Quantity Under Varying Field Conditions," Soil Sci. Soc. Amer. 104:227-233 (1967).

28. Tanji, K. K., J. D. Doneen, G. V. Ferry, and R. S. Ayers. "Computer Simulation Analysis on Reclamation of Salt-Affected Soils in San Joaquin Valley, CA," Soil Sci. Soc. Amer. Proc. 36:127-133 (1972).

29. Burns, I. G. "A Model for Predicting the Redistribution of Salts Applied to Fallow Soils After Excess Rainfall or Evaporation," J. Soil Sci. 25:165-178 (1974).

30. Addiscott, T. M. "A Simple Computer Model for Leaching in Structured Soils," J. Soil Science 28:554-563 (1977).

31. Dagan, G., and E. Bresler. "Solute Dispersion in Unsaturated Heterogeneous Soil at Field Scale. I. Theory," Soil Sci. Amer. J. 43:461-466 (1979).

32. Amoozegar-Fard, A., D. R. Nielsen, and A. W. Warrick. "Soil Solute Concentration Distributions for Spatially-Varying Pore Water Velocities and Apparent Diffusion Coefficients," Soil Sci. Soc. Amer. J. 46:3-9 (1982).

33. Jury, W. A. "Simulation of Solute Transport Using a Transfer Function Model," Water Resour. Res. 18:363-368 (1982).

34. Jury, W. A., L. A. Stolzy, and P. Shouse. "A Field Test of the Transfer Function Model for Predicting Solute Transport," Water Resour. Res. 18:369-375 (1982).

35. van Genuchten, M. Th., and W. J. Alves. "Analytical Solutions of the One-Dimensional Convective Dispersive Solute Transport Equation," Bulletin 1661, U.S. Department of Agriculture (1982).

36. Enfield, C. G., R. F. Carsel, S. Z. Cohen, T. Phan, and D. M. Walters. "Approximating Pollutant Transport to Ground Water," Ground Water 20:711-722 (1982).

37. Lorber, M. Personal communication (1985).

38. Jones, R. L., and R. C. Back. "Monitoring Aldicarb Residues in Florida Soil and Water," Environ. Toxicol. Chem. 3:9-20 (1984).

39. Melancon, S. M., J. E. Pollard, and S. C. Hern. "Evaluation of SESOIL, PRZM, and PESTAN in a Laboratory Column Leaching Experiment," Environ. Toxicol. Chem. 5: (In Press, 1986).

40. Bonazountas, M., and J. Wagner. "SESOIL: A Seasonal Soil Compartment Model" (Cambridge, MA: Arthur D. Little, Inc., 1984).

41. Bonazountas, M., and J. Wagner. "Pollutant Transport in Soils via 'SESOIL,'" Presented at ASCE National Conference of Environmental Engineering, Minneapolis, MN July 14-16, 1982.

42. Eagelson, P. S. "Climate, Soil, and Vegetation," Water Resour. Res. 14:1-7 (1978).

43. Wischmeier, W. H., and D. D. Smith. "Predicting Rainfall Erosion Losses from Cropland - A Guide to Conservation Planning," Agricultural Handbook 537, U.S. Department of Agriculture (1978).

44. Foster, G. R., L. J. Lane, J. D. Nowlin, J. M. Laflen, and R. A. Young. "A Model to Estimate Sediment Yield from Field-Sized Areas: Development of Model, Purdue Agricultural Experiment Station," Purdue Journal No. 7781 (1980).

45. Knisel, W., Ed. "CREAMS: A Field-Scale Model for Chemicals, Runoff, and Erosion from Agricultural Management Systems," Conservation Research Report No. 26, U.S. Department of Agriculture (1980).

46. Bonazountas, M., J. Wagner, and B. Goodwin. "Evaluation of Seasonal Soil/ Groundwater Pollutant Pathways" (Cambridge, MA: Arthur D. Little, Inc., 1982).

47. Wagner, J., M. Bonazountas, and M. Alsterberg. "Potential Fate of Buried Halogenated Solvents via SESOIL (Cambridge, MA: Arthur D. Little, Inc., 1983).

48. Hetrick, D. M. "Simulation of the Hydrologic Cycle for Watersheds," Proceedings for the Applied Simulation and Modeling Conference, San Francisco, CA (1984).

49. Hetrick, D. M., J. T. Holdeman, and R. J. Luxmore. "AGTHEM: Documentation of Modifications to the Terrestrial Ecosystem Model (TEHM) for Agricultural Applications," ORNL/TM-7856 (Oak Ridge, TN: Oak Ridge National Laboratory, 1982).

50. Hetrick, D. M., and L. M. McDonald-Boyer. "User's Manual for TOX-SCREEN: Multimedia Screening-Level Program for Assessing Potential Fate of Chemicals Released to the Environment," ORNL-6041 (Oak Ridge, TN: Oak Ridge National laboratory) and EPA-560/83-024 (Washington, DC: U.S. EPA, 1984).

51. Bicknell, B. R., S. H. Boutwell, and D. B. Watson. "Testing and Evaluation of the TOX-SCREEN Model" (Palo Alto, CA: Anderson-Nichols and Co., 1984).

52. Kincaid, C. T., J. R. Morery, S. B. Yabusaki, A. R. Felmy, and J. E. Rogers. "Geohydrochemical Models for Solute Migration. Volume 2: Preliminary Evaluation of Selected Computer Codes for Modeling Aqueous Solutions and Solute Migration in Soils and Geologic Media," EA-3417 (Palo Alto, CA: Electric Power Research Institute, 1984).

53. Watson, D. B., and S. M. Brown. "Testing and Evaluation of the SESOIL Model" (Palo Alto, CA: Anderson-Nichols and Co., Inc., 1984).

54. Nold, A. Personal communication (1985).

55. Bonazountas, M. Personal communication (1985).

56. Carsel, R. F., C. N. Smith, L. A. Mulkey, J. D. Dean, and P. Jowise. "User's Manual for the Pesticide Root Zone Model (PRZM): Release 1," EPA-600/3-84-109 (Athens, GA: U.S. EPA, 1984).

57. Williams, J. R. "Sediment Yield Prediction with Universal Equation using Runoff Energy Factor," in Present and Prospective Technology for Predicting Sediment Yields and Sources, AARS-S-40, U.S. Department of Agriculture (1975).

58. Carsel, R. F., L. A. Mulkey, M. N. Lorber, and L. B. Baskin. "The Pesticide Root Zone Model (PRZM): A Procedure for Evaluating Pesticide Leaching Threats to Groundwater," Ecol. Modeling 30:49-69 (1985).

59. Jones, R. L. "Movement and Degradation of Aldicarb Residues in Soil and Groundwater," Presented at the Environmental Toxicology and Chemistry Conference on Multidisciplinary Approaches to Environmental Problems, Crystal City, VA (1983).

60. Carsel R. F., W. B. Nixon, and L. G. Balentine. "Comparison of Pesticide Root Zone Model Predictions with Observed Concentrations for the Tobacco Pesticide Metalaxyl in Unsaturated Zone Soils," Environ. Toxicol. Chem. 5:345-353 (1986).

61. Carsel, R. Personal communication (1984).

62. Dean, J. D., P. P. Jowise, and A. S. Donigian, Jr. "Leaching Evaluation of Agricultural Chemicals (LEACH) Handbook," EPA-600/3-84-068 (Athens, GA: U.S. EPA, 1984).

63. Dean, J. D., A. S. Donigian, Jr., and J. E. Rafferty. "Development of a Mathematical Model to Evaluate Pesticide Exposure to Birds in Agricultural Ecosystems," (Palo Alto, CA: Anderson-Nichols and Co., Inc., 1984).

64. Donigian, A. S. Jr., and R. F. Carsel. "Impact of Conservation Tillage on Pesticide Concentrations in Ground Water, Surface Water, and Air. Draft Report" (Washington, DC: U.S. EPA, 1985).

65. Brown, S. M., and S. H. Boutwell. "PCB Spill Exposure Assessment Methodology," (Palo Alto, CA: Anderson-Nichols and Co., Inc., 1985).

66. Dean, J. D., and D. F. Atwood. "Exposure Assessment Modeling for Aldicarb in Florida" (Athens, GA: U.S. EPA, 1985).

67. Donigian, A. S. Jr., S. M. Brown, and S. B. Yabusaki. "Ground-Water Modeling of Selected Hydrogeologic Settings to Determine Leachate Fate and Migration from Waste Facilities" (Washington, DC: U.S. EPA, 1983).

68. Leonard, R. A., and V. A. Ferreira. "CREAMS2 - The Nutrient and Pesticide Models," Proceedings of the Natural Resources Modeling Symposium. U.S. Department of Agriculture. (In Press, 1984).

69. Leonard, R. A., and W. G. Knisel, Jr. "Model Selection for Nonpoint Source Pollution and Resource Conservation," Proceedings of the International Conference on Agriculture and Environment, Venice, Italy (1984).

70. Schroeder, P. R., J. M. Morgan, T. M. Walbki, and A. C. Gibson. "The Hydrologic Evaluation of Landfill Performance (HELP) Model, Volume I. User's Guide for Version I," EPA/530-SW-84-009. (Washington DC: U.S. EPA, 1984).

71. Bicknell, B. R. "Modeling Chemical Emissions from Lagoons and Landfills, Final Report" (Athens, GA: U.S. EPA, 1984).

72. Oster, C. A. "Review of Ground Water Flow and Transport Models in the Unsaturated Zone," PNL-4427, NUREG/CR-2917 (Richland, WA: Battelle, Pacific Northwest Laboratory, 1982).

73. Kincaid, C. T., J. R. Morery, and J. E. Rogers. "Geohydrochemical Models for Solute Migration, Volume 1: Process Description and Computer Code Selection," AEA-3417 (Palo Alto, CA: Electric Power Research Institute, 1984).

CHAPTER 2

TRANSPORT MECHANISMS AND LOSS PATHWAYS FOR CHEMICALS IN SOIL

W. A. Jury
University of California-Riverside, Riverside, California

R. L. Valentine
University of Iowa, Iowa City, Iowa

TRANSPORT MECHANISMS

Chemicals are transported through soil principally by three mechanisms: mass flow of dissolved chemical within moving solution, liquid diffusion within soil solution, and gaseous diffusion within soil air voids. The first mechanism, mass flow, refers to the passive transport of dissolved solute within moving soil water which is approximated as the product of the volume flux of water times the dissolved solute concentration. Liquid diffusion refers to the transport of the dissolved solutes within solution by intermolecular collisions which move the solutes from regions of the higher solute density to lower solute density. The diffusion flux is written as the product of the density or concentration gradient and a coefficient of proportionality called the soil liquid diffusion coefficient. Similarly, chemical vapor molecules in the soil air spaces also undergo molecular collisions and spread out by vapor diffusion which is expressed as the product of the vapor density or concentration gradient and a proportionality coefficient called the soil vapor diffusion coefficient.

Since the soil water flux is represented as a continuous volume-averaged quantity over many pores, the individual complicated water flow paths around the soil grains are replaced by an equivalent one-dimensional water flow. When this one-dimensional water flow is multiplied by the dissolved solute concentration, the resulting mass flux does not take into account the extra solute spreading which occurs by three-dimensional mass flow at the pore scale in the actual system and is not represented in the volume-averaged mathematical treatment. This apparent solute diffusion arising from the mass flux effects which are obscured by mathematical volume averaging is called hydrodynamic dispersion. Under certain conditions, this dispersion effect is mathematically equivalent to liquid diffusion and may be included as a diffusion-like transport mechanism by using an effective liquid diffusion-dispersion coefficient to account for both dispersion and diffusion.

LOSS PATHWAYS FOR CHEMICALS IN SOIL

Chemicals added to a soil profile from the surface may leave the zone of incorporation by one of three loss pathways. The first pathway, known as leaching, takes place principally by mass flow and refers to the downward movement of dissolved chemical. The second loss pathway, volatilization, refers to the loss of chemical vapor to the atmosphere through the soil surface. The final loss pathway, degradation, refers to the biological or chemical transformation of the chemical to a different form with properties distinct from those of the chemical prior to transformation. Each of these loss pathways will be discussed in greater detail below.

PARAMETERS INFLUENCING TRANSPORT PROCESSES AND LOSS PATHWAYS

Table 2-1 summmarizes the significant soil, environmental, and managerial parameters influencing the four transport processes and atmospheric volatilization. Since mass flow is the product of water flux and dissolved chemical concentration, it is most strongly influenced by those parameters having a direct influence on these two variables. Thus, increases in the amount of applied water cause increases in water flux and hence in mass flow for a given amount of chemical. Furthermore, in structured soils, increases in water application intensity for a given amount of applied water may increase the probability of macropore flow (flow through cracks or channels) which can move small amounts of chemical great distances.

For a given total amount of chemical, the fraction which exists in the dissolved phase and hence is available for mass flow is strongly influenced by the adsorptive properties of the soil matrix. For nonpolar organic molecules, most of the adsorption takes place on organic matter surfaces. Thus, there is a strong relationship between organic matter or organic carbon content and total adsorption. For positively charged species such as inorganic cations and a few pesticides like paraquat and diquat, the negatively charged clay mineral surfaces will be the principal adsorption sites, unless the soil is strongly acidic; if the soil is strongly acidic, negatively charged species may be adsorbed as well. Unless the adsorption isotherm is strongly nonlinear, increasing the total concentration of the chemical should give a nearly proportional increase in the solution concentration and thus proportionally affect mass flow.

Table 2-1. Summary of Significant Process-Parameter Relationships in Chemical Transport and Loss to the Atmosphere.

Process	Parameter	Influence
1. Mass Flow	a. Amount of applied water	Mass flow of dissolved chemical is directly proportional to water input.
	b. Adsorption site density	Adsorption decreases dissolved chemical concentration and decreases mass flow. Nonpolar organic chemicals bind primarily to organic matter, whereas cationic (positively charged) species adsorb to negatively-charged mineral surfaces. Anionic (negatively charged) species may also bind to soil minerals for soils of low pH.
	c. Chemical concentration	For chemicals whose water solubility is not exceeded, increased chemical concentration will increase dissolved chemical concentration and increase mass flow. The increase may be greater than proportional if the adsorption isotherm of the chemical is nonlinear.
	d. Water application intensity	For a given amount of water application, increased water application rate may increase mass flow by decreasing rate-limited adsorption and may produce increased macropore (pore bypass) flow thereby causing deeper penetration of part of the chemical mass.
	e. Saturated hydraulic conductivity K	Under ponded conditions, mass flow is proportional to saturated hydraulic conductivity. For subsaturated water application, control is regulated by the surface application, and the hydraulic conductivity has no

(continued)

Table 2-1. (continued)

Process	Parameter	Influence
		direct effect on mass flow. However, soils with large K generally retain less water, and penetration is deeper for a given water application.
	f. Temperature T	For many chemicals, increasing temperature decreases adsorption and increases mass flow. Saturated hydraulic conductivity increases with temperature.
2. Vapor Diffusion	a. Water content θ	Vapor diffusion decreases strongly with increasing water content. A frequently used model assumes that the soil vapor diffusion coefficient is proportional to $(\emptyset-\theta)^{10/3}$, where $\emptyset$ is porosity.
	b. Adsorption site density	Adsorption decreases gaseous chemical concentration and decreases vapor diffusion. Most volatile adsorbed chemicals are nonpolar and adsorb primarily to organic matter.
	c. Chemical concentration	For chemicals whose vapor density is not saturated, increasing chemical concentration will increase vapor density and increase vapor diffusion. The increase may be greater than proportional if the chemical vapor adsorption isotherm is nonlinear.
	d. Temperature	Increasing temperature significantly increases vapor density for a given amount of chemical in soil, thereby increasing vapor diffusion. The vapor diffusion

(continued)

Table 2-1. (continued)

Process	Parameter	Influence
		coefficient increases nonlinearly with increasing temperature, proportional to $T^{1.75}$ (Kelvin).
	e. Henry's constant K_H	Henry's constant (ratio of saturated vapor density to solubility) is an index of the partitioning of a chemical between dissolved and gaseous phases. The larger the K_H, the more likely the chemical is to move by vapor diffusion as opposed to liquid diffusion.
3. Liquid Diffusion	a. Water content θ	Liquid diffusion increases strongly with increasing water content. A frequently used model assumes that the soil liquid diffusion coefficient is proportional to $\theta^{10/3}$.
	b. Adsorption site density	Adsorption decreases dissolved chemical concentration and decreases liquid diffusion (see 1b above).
	c. Chemical concentration	For chemicals whose solubility is not exceeded, increasing chemical concentration will increase dissolved chemical concentration, thereby increasing liquid diffusion. The increase may be greater than proportional if the adsorption isotherm is nonlinear.
	d. Henry's constant K_H	Chemicals with low K_H will primarily diffuse in the liquid phase as opposed to the vapor phase.

(continued)

Table 2-1. (continued)

Process	Parameter	Influence
	e. Temperature	Liquid diffusion increases with increasing temperature because adsorption generally decreases. The liquid diffusion coefficient increases modestly with increasing T.
4. Hydrodynamic Dispersion	a. Water application	The dispersion coefficient is usually assumed to be proportional to water velocity which is proportional to water application rate.
	b. Scale of problem	Dispersion is a pseudo-mechanism resulting from representing mass flow by an average one-dimensional model while neglecting convective mixing which is represented as dispersion. The dispersion coefficient is thus dependent on the size of the simulation and may increase by orders of magnitude between lab and field.
5. Volatilization to the Atmosphere	a. Henry's constant K_H	Chemicals with large K_H are generally more volatile than those with small K_H and tend to have volatilization rates which are limited more by soil conditions than atmospheric conditions.
	b. Chemical concentration	For chemicals whose vapor density is not saturated, increasing concentration will increase volatilization (see 2c).
	c. Adsorption site density	Volatilization decreases as adsorption increases (see 2b).

(continued)

Table 2-1. (continued)

Process	Parameter	Influence
	d. Temperature	Volatilization increases significantly as temperature increases (see 2d).
	e. Water content θ	Decreasing water content increases vapor diffusion and volatilization (see 2a). If the surface dries out completely, adsorption increases dramatically, and volatilization may cease until surface is rewet.
	f. Wind speed	Increased wind speed improves mixing with the atmosphere and can increase volatilization especially of chemicals with low K_H.
	g. Water evaporation	Water evaporation aids in transporting chemical to the surface thus increasing volatilization especially of weakly adsorbed chemicals.

When the soil is unsaturated, mass flow is principally regulated by the amount of water applied at the surface. However, under saturated conditions when the soil surface is ponded, the saturated hydraulic conductivity of the soil will have a dominant effect on the amount of water entering the soil and hence on mass flow. Under such conditions, it is also more likely that adsorption will be rate-limited and less effective, and the extent of chemical transport by mass flow will be greater for a given amount of applied water than when the soil is unsaturated.

Temperature principally influences mass flow by increasing the total fraction of chemical mass which is present in the dissolved phase. Increasing temperature also lowers the viscosity of water which modestly increases the saturated hydraulic conductivity (approximately 25 percent between 20°C and 30°C).

The principal factor influencing the amount of transport by vapor diffusion for a given chemical is the fraction of the total chemical concentration which is present in the vapor phase. In assessing this tendency, two indices are important: adsorption and the Henry's constant K_H. As a general rule, increasing the total adsorption of the chemical to soil lowers the vapor concentration, and decreasing the value of Henry's constant K_H lowers the vapor concentration. Chemicals with Henry's constants lower than 10^{-4} in dimensionless units tend to be found principally in the dissolved as opposed to the vapor phase [1]. For chemicals with an appreciable vapor concentration, soil water content strongly influences the amount of transport by vapor diffusion. One frequently-used model, called the Millington and Quirk tortuosity model, expresses the ratio of the soil to air diffusion coefficient as the 10/3 power of the volumetric air content [2].

For a given amount of chemical present in the soil, increasing temperature can strongly enhance the relative amount of movement by vapor diffusion by increasing the Henry's constant and also by raising the value of the vapor diffusion coefficient.

The extent of solute transport by liquid diffusion in soil is moderated by the amount of dissolved chemical and the space available for flow. Increases in soil adsorption tend to decrease the concentration in the dissolved phase and hence decrease liquid diffusion. Also, chemicals with low Henry's constant tend to be found primarily in the liquid phase and hence have a higher affinity for liquid as opposed to vapor diffusion. Increases in water content strongly increase the tendency for liquid diffusion by increasing the accessible flow volume. The Millington and Quirk model, also applicable to liquid diffusion, predicts that the soil liquid diffusion coefficient is proportional to the 10/3 power of the volumetric water content. As is the case with vapor diffusion, increases in the total amount of chemical applied to soil tend to increase the concentration in the dissolved phase and hence proportionally increase the amount of movement by liquid diffusion. Contrary to vapor diffusion, however, temperature increases only modestly increase the amount of movement by liquid diffusion.

Hydrodynamic dispersion is really a form of mass flow and hence is governed by the principles discussed above for mass flow. However, the most dominant influence on hydrodynamic dispersion is the scale over which the water flux is averaged. Hence, should one attempt to use a one-dimensional chemical transport model over a large area such as a field, one would have to use a very large value of hydrodynamic dispersion coefficient to take into account the chemical transport by variable water velocities not included in the one-dimensional average for the mass flux.

Volatilization of chemical vapor to the atmosphere is an extremely complex process controlled by soil, chemical, and atmospheric influences. As a general rule, however, chemicals with a large value of Henry's constant K_H are more volatile than those with low K_H and tend to be relatively independent of atmospheric conditions. Since volatilization takes place in the vapor phase, processes which limit vapor diffusion, such as increased adsorption and increased water content, can decrease the amount of volatilization. Conversely, processes which increase vapor diffusion such as increased temperature and increased total concentration will increase the amount of volatilization. Water evaporation can also increase volatilization as upward water flow tends to bring chemical to the surface to replace that which is lost by volatilization. This enhancement of volatilization flux is most pronounced when the chemical is weakly adsorbed. Finally, increases in wind speed increase mixing with the atmosphere at the soil surface and increase the amount of volatilization. This increase is most dramatic for chemicals which have a low Henry's constant and may have only minimal volatilization under stagnant surface conditions.

MEASUREMENT OF KEY PARAMETERS

Table 2-2 summarizes some of the published methods of measurement of various model parameters or soil properties which are relevant to modeling and validation. The first set of properties, classified as static soil properties, are each measured from intact or disturbed samples in the laboratory by standard methods. Also, most of these properties tend to have only modest amounts of variation across large soil areas and thus could be characterized from a few replicates (see Chapter 11).

Conversely, the water transport and retention functions tend to have higher coefficients of variation and different properties when measured under field conditions than when the measurements are made in the laboratory on intact or disturbed soil cores.

The third group, called basic chemical properties, contains all pure chemical properties which are measured in the absence of soil. In many cases, values of the parameters may already be available in the literature (see Table 2-3). For those situations where values are not available, the references provide the accepted protocol for measurement.

Table 2-2. Methods of Measurement of Model Parameters or Soil Properties Relevant to Modeling and Validation.

Parameter	Locale of Measurement	Methodology	Reference
1. Static Soil Properties			
Porosity	Laboratory	Water content at zero suction on undisturbed cores	[3]
Bulk Density	Laboratory	Coring into known volume or intact clod of soil	[4]
Particle size (% Sand, etc.)	Laboratory	Hydrometer or pipette method after sieving	[5]
Specific surface area	Laboratory	Soil adsorption of gases or polar liquids	[6]
Organic carbon	Laboratory	Walkley-Black chromic acid titration method	[7]
Cation exchange capacity	Laboratory	Sodium saturation followed by ammonium acetate replacement	[8]
pH	Laboratory	pH meter reading of paste extract or soil solution	[8]
2. Water Transport and Retention Functions			
Saturated hydraulic conductivity	Laboratory	Ponding of soil cores (steady or falling head method)	[9]
	Field	Steady state infiltration while monitoring pressure head	[10]
	Field	Air entry permeameter	[11]
Matric potential-water content function	Laboratory	Hanging water table and pressure plate	[12]
	Field	Simultaneous tensiometer-neutron probe measurements	[13]

(continued)

Table 2-2. (continued)

Parameter	Locale of Measurement	Methodology	Reference
Unsaturated hydraulic conductivity	Laboratory	Steady flow in soil columns	[8]
	Field	Instantaneous profile method	[14]
	Field	Unit gradient methods	[15]
	Field	Air entry permeameter	[11]
Solute velocity and dispersion coefficient	Laboratory	Parameter fitting to column breakthrough curve	[16]
	Field	Parameter fitting to soil core or solution sampler data	[17]
Water flux	Field	Indirectly, by measuring solute travel time	[18]
Volatilization Flux	Field	Atmospheric monitoring	[19]
3. Basic Chemical Properties			
Vapor pressure	Laboratory	Gas saturation	[20]
Water solubility	Laboratory	Dilution of solute	[21]
Henry's constant K_H	Derived	Ratio of vapor pressure (or density) to solubility	
Vapor diffusion coefficient in air	Literature	Estimate from similar compounds tabulated	[22]
Liquid diffusion coefficient in water	Literature	Estimate from similar compounds tabulated	[23]
Octanol-water partition coefficient	Laboratory	Equilibration with octanol-water mix	[24]

(continued)

Table 2-2. (continued)

Parameter	Locale of Measurement	Methodology	Reference
4. Time Dependent Parameters Requiring Monitoring			
Water content	Field to Laboratory	Gravimetric determination from soil core	[25]
	Field	Neutron probe	[25]
Solute concentration	Field	Solution samplers and soil cores	[8]
Potential evaporation	Field	Various correlations or energy balance methods	[26]
Volatilization boundary layer thickness	Derived	Calculate from water evaporation rate	[27]
5. Soil Adsorption Parameters			
Distribution coefficient	Laboratory	Batch adsorption to equilibrium	[28]
Freundlich isotherm	Laboratory	Batch adsorption to equilibrium	[28]
Organic carbon partition coefficient	Derived	Ratio of distribution coefficient to organic carbon fraction	
6. Tortuosity Functions			
Vapor diffusion tortuosity	Laboratory	(Both functions have been found to fit model of Millington and Quirk)	[27]
Liquid diffusion tortuosity	Laboratory		

Table 2-3. Sources of Measured Values for Various Chemical Properties.

Property	References
Vapor diffusion coefficient in air	[22]
Liquid diffusion coefficient in water	[23]
Organic carbon partition coefficients	[1,28,29,30,31]
Octanol-water partition coefficients	[1,24,29,31]
Vapor density or vapor pressure	[1,32]
Water solubility	[1,29,32]
Henry's constant	[1,32]
Degradation rate constant	[1,31,33,34]

The fourth group consists of those time-dependent parameters which require periodic monitoring during an experiment. The references provide standard methods for field measurement of these variables.

Soil adsorption parameters (group five), must be measured in the laboratory, although adsorption parameters may also be inferred from soil column breakthrough curves obtained on undisturbed cores originally taken from the field [35]. There is no method available for measuring soil adsorption _in situ_.

More complex adsorption relationships, which are discussed in Section II of this report, must be obtained in laboratory equilibrium or transport experiments by curve fitting. It is difficult to make these laboratory experiments representative of the field environment.

The tortuosity functions in parameter group six which describe the reduction in diffusion coefficient when soil is present can only be obtained in laboratory experiments in which the volumetric water content is known. Substantial experimentation over the last two decades has shown the Millington and Quirk model to be an accurate representation of the influence of water content on vapor and liquid diffusion. Thus, use of this model is recommended in lieu of experimental measurements.

TABULATIONS OF MEASURED VALUES

Table 2-3 provides references for tabulated values of key parameters for a large number of chemicals. These values frequently vary significantly between studies and should be used only when experimentation is not feasible. Further detailed descriptions of the transport mechanisms and chemical pathways are found in Chapters 6-8; significant mathematical derivations of equations related to these processes are in Appendix A.

TRANSFORMATION PROCESSES

Chemicals added to a soil may be transformed to products having properties distinct from those of the chemical prior to transformation. While all processes leading to structural changes in a chemical occur as the result of chemical reactions, these processes can be categorized as being either biological or nonbiological transformations, depending on the role of microorganisms. Nonbiological transformations can be classified further into chemical and photochemical transformations depending on the role of light. In general, the transformation products of nonbiological and biological processes may be similar and encompass a wide variety of specific compounds characteristic of many kinds of chemical reactions. For example, hydrolysis and oxidation-reduction reactions are two general classes of reactions which typically occur by biological and nonbiological processes. Detailed discussions of biological and nonbiological transformation processes can be found in Chapters 9 and 10.

BIOTRANSFORMATION

Biological transformations occur as the result of the metabolic activity of microorganisms, primarily attached bacteria, actinomycetes, and fungi, through the action of enzymes which catalyze chemical reactions. Not all chemicals are transformed by all organisms. Many chemicals are degraded only by a select group of organisms. Additionally, the transformation may involve one or more species and enzymes and form several products through complex reaction pathways. These reactions usually lead to the production of energy or some essential nutrient. However, some chemicals are also transformed even though the transformation does not promote growth. Biotransformation of this sort has been termed *cometabolism* and can occur when other substances are present which support growth.

NONBIOLOGICAL TRANSFORMATION

Certain classes of organic compounds may undergo chemical *hydrolysis*. These reactions result in the net exchange of a hydroxyl group, OH^-, with some other group, including halides, alcohols, amines, and sulfur- and phosphorus containing moieties. Many compounds do not undergo hydrolysis reactions. However, likely candidates can be ascertained from knowledge of their structure. Additionally, hydrolysis rates may vary greatly even among compounds of similar structure. Many chemicals may also be oxidized or reduced by substances in the soil. *Oxidation-reduction* reactions constitute a very general class of reactions characterized by a change in the oxidation state of a chemical. Many specific changes in the identity of a chemical can be classified as having occurred via redox reactions.

Photochemical transformations occur as the result of light being absorbed by the specific chemical of interest or by some other substance which then reacts with it. Photochemical transformations include a variety of common reactions such as oxidation, reduction, substitution, elimination, etc. Since light does not penetrate very deeply into the soil, photochemical transformations are potentially important only at the surface of the soil.

FACTORS INFLUENCING BIOLOGICAL AND NONBIOLOGICAL TRANSFORMATION

The biotransformation of a chemical depends primarily on factors which govern the nature of the microbial population, including size and composition, and the availability of the chemical to attack by the organisms or their enzymes. The nonbiological transformation of a chemical is determined primarily by factors which govern the availability of reactants and the nature of the specific chemical in the soil.

Many of the factors which affect biological transformation also affect nonbiological transformation processes. Furthermore, many of these factors are highly interrelated. Any factor which affects sorption can be expected to have an effect on biological and nonbiological transformation processes. Processes occurring at the soil-water interface may differ both in rate and in kind from those occurring in solution. Table 2-4 lists some of the more important factors that influence transformation processes.

Microbial and chemical concentration are the two most common parameters which are incorporated into mathematical expressions describing the rate of biotransformation. It is axiomatic that the rate of biotransformation should increase with an increase in the population of actively degrading microorganisms. Likewise, the rate should also increase with increasing chemical concentration (at least over some range in concentration) unless the chemical is toxic to the degrading organisms. The rates of chemical reactions are expected to be a function of those compounds reacting in the rate limiting step(s).

Increases in temperature usually cause increased biotransformation rates at least over the range in temperatures tolerated by the degrading organisms. Temperature may also affect species composition and the availability of the chemical through sorption. Temperature-dependent biotransformation rate constants have been incorporated into models often using some relationship that can be rationalized in terms of the Arrhenius equation typically used to relate chemical reaction rates to temperature.

Oxygen plays a major role in determining the types of organisms present and the metabolic pathways used to degrade chemicals. Some chemicals are degraded only by aerobic organisms which require oxygen as the terminal electron acceptor in their metabolic pathway. Others are degraded by anaerobic organisms which exist only in the

Table 2-4. Major Factors Influencing Biological and Nonbiological Transformations.

Factor	Primary Influence	
	Biological	Nonbiological
1. Microbial concentration	Rates increase with increasing active population.	Not applicable (NA)
2. Chemical concentration	Rates usually increase unless chemical is toxic.	Rates may increase but depends on reaction mechanism/process.
3. Temperature	Rates increase with relatively small increases in temperature over range tolerated by organisms.	Rates increase with increases in temperature.
4. Oxygen	Determines types of organisms and degradation pathways (aerobic vs. anaerobic).	Potential oxidant.
5. Nutrients	Needed for microbial metabolism.	NA
6. Inorganic and organic composition	Organic carbon may support microbial population; affects bioavailability.	Potential reactants and catalysts; affects availability.
7. pH	Affects population composition; affects bioavailability.	Major effect on hydrolysis, probable effect on most chemical reactions; affects availability.
8. Soil water content	Needed for growth; affects bioavailability; may limit oxygen content causing anaerobic system.	Needed for hydrolysis; may affect soil-water interface pH; affects chemical availability; may limit oxidation by oxygen.
9. Acclimation	Possible time-lag in degradation while some enzymes induced.	NA
10. Light	NA	Increases transformation of certain chemicals at soil surface.

absence of oxygen. Some organisms can grow both in the absence and presence of oxygen by switching metabolic pathways. It is generally under aerobic conditions that a chemical can be mineralized, i.e., degraded to carbon dioxide and other inorganic products. Frequently, the products of anaerobic degradation are very similar to the original parent compound. Oxygen may be an important oxidant in the nonbiological oxidation of a chemical.

A variety of substances are needed for growth of an organism and must be available in order for organisms to degrade a chemical. For example, nitrogen, phosphorus, and carbon are universally needed by all organisms along with many trace inorganic substances. These are readily available in most soils. The primary source of carbon available to the degrading population may be the naturally occurring soil organic carbon, not the added chemical. Microbial concentrations are frequently correlated with soil organic carbon content. Certain anaerobic organisms require inorganic compounds such as sulfate and nitrate which are used as electron acceptors in the degradation mechanism. Naturally occurring organic and inorganic substances may act as oxidants or reductants in nonbiological transformation pathways. Some metals may also act as catalysts in these processes.

The hydrogen ion content of the soil as measured by pH also affects both the types of organisms present and the rate at which chemicals may be degraded. Some organisms can exist only over a narrow range in pH. The rate of hydrolysis is frequently directly related to pH, and pH may affect the rate of redox reactions.

Water is required by all microorganisms and serves as the media through which a chemical is made available to the attached microorganisms. Microbial activity has been correlated over limited ranges to moisture content. Moisture content affects sorption and is therefore expected to affect the rates of transformation of chemicals in the soil. Soil moisture content may also govern the oxygen content in the soil by limiting the rate at which it diffuses into the soil. Hence a waterlogged soil may become anaerobic. The pH of the soil-water interface is frequently a function of moisture content.

Many enzymes used to degrade a particular chemical must be first induced after the chemical comes into contact with the microorganism. This can lead to long acclimation periods over which little chemical is transformed.

MEASUREMENT OF KEY PARAMETERS

The mathematical treatment of transformation is frequently based on an approach which "lumps" together all processes due to the difficulty in distinguishing between biological and nonbiological transformations without extensive studies. Such models usually consider the transformation of a chemical as a single first order degradative process and are inherently the most site specific.

Models which treat each transformation mechanism independently are potentially applicable to a wider variety of field conditions and therefore may be of greater usefulness.

Biotransformation rate expressions can be rationalized based on the work of Monod and Michaelis-Menton. First order rate expressions have been frequently used to model biotransformation. The first order rate constant is an implicit function of the concentration of the actively degrading microbial population present in the soil sample used to obtain it. Second order models which are an explicit function of both chemical concentration and concentration of the actively degrading microbial population are potentially less site specific. Hydrolysis rate expressions are expected to be first order in chemical concentration. However, rate expressions for other chemical processes occurring in soil have not been developed from first principles. Nonbiological transformations are frequently modeled as first order processes for lack of some better rate expression. Regardless of the seemingly detailed considerations given to the evaluation of rate constants, they should be used with caution because of the many unknown and possibly variable factors which affect them.

Separate biotransformation and nonbiological transformation rate coefficients cannot easily be evaluated under field conditions because of concurrent abiotic and biological processes which lead to chemical loss. Laboratory studies using sterile controls must be conducted to resolve biological from nonbiological transformation processes. Most commonly used rate coefficients are based on the rate of loss of the parent chemical concentration in the soil without differentiating sorbed chemical from that in true solution or considering the extent of change in the identity of the parent chemical. These coefficients can be determined under aerobic and anaerobic conditions. The potential importance of photochemical transformations at the soil surface can be evaluated using natural and artificial light. In general, methodologies for the evaluation of rate constants and rate parameters in soil systems are not well developed. Table 2-5 lists some references that should be of use in evaluating rate constants and parameters.

Table 2-5. General Approaches and Methodologies for the Determination of Transformation Parameters.

Methodology/Transformation Process	Reference
1. General field and laboratory testing	[36,37,38,39,40,41,42]
2. Use of microcosms	[43]
3. Subsurface sampling	[44,45,46,47,48]
4. Determination of microbial activity	[49,50,51,52,53,54,55,56,57]
5. Biological	[39,40,58]
6. Chemical	[39,40,52,59,60,61,62]
7. Photochemical	[39,40,63,64,65,66,67]

REFERENCES

1. Jury, W. A., W. F. Spencer, and W. J. Farmer. "Behavior Assessment Model for Trace Organics in Soil. III. Application of Screening Model," J. Environ. Qual. 4:573-578 (1984).

2. Millington, R. J. and J. M. Quirk. "Permeability of Porous Solids," Trans. Faraday Soc. 57:1200-1207 (1961).

3. Foth, H. D. Fundamentals of Soil Science, 6th Ed. (New York: John Wiley & Sons, 1978).

4. Blake, G. R. "Bulk Density," in Methods of Soil Analysis, C. A. Black, Ed. Monograph 9 (Madison, WI: American Society of Agronomy, 1965).

5. Day, P. R. "Particle Fractionation and Particle Size Analysis," in Methods of Soil Analysis, C. A. Black, Ed. Monograph 9 (Madison, WI: American Society of Agronomy, 1965).

6. Mortland, M. M., and W. D. Kemper. "Specific Surface," in Methods of Soil Analysis, C. A. Black, Ed. Monograph 9 (Madison, WI: American Society of Agronomy, 1965).

7. Allison, L. E. "Organic Carbon," in Methods of Soil Analysis, C. A. Black, Ed. Monograph 9 (Madison, WI: American Society of Agronomy, 1965).

8. Richards, L. A. "Diagnosis and Improvement of Saline and Alkali Soils," Agriculture Handbook 60, U.S. Department of Agriculture (1954).

9. Klute, A. "Laboratory Methods of Hydraulic Conductivity of a Saturated Soil," in Methods of Soil Analysis, C. A. Black, Ed. Monograph 9 (Madison, WI: American Society of Agronomy, 1965).

10. Nielsen, D. R., J. W. Biggar, and K. T. Erh. "Spatial Variability of Field-Measured Soil-Water Properties," Hilgardia 42:215-259 (1973).

11. Russo, D., and E. Bresler. "Field Determination of Soil Hydraulic Properties for Statistical Analyses," Soil Sci. Soc. Amer. J. 44:696-702 (1980).

12. Taylor, S. A., and G. L. Ashcroft. Physical Edaphology (San Francisco, CA: W. H. Freeman and Co., 1972).

13. Jury, W. A., H. Frenkel, H. Fluhler, D. Devitt, and L. H. Stolzy. "Use of Saline Irrigation Waters and Minimal Leaching for Crop Production," Hilgardia 46:169-192 (1978).

14. Rose, C. W., W. R. Stern, and J. E. Drummond. "Determination of Hydraulic Conductivity in situ," Aust. J. Soil Res. 3:1-9 (1965).

15. Libardi, P. L., K. Reichart, D. R. Nielsen, and J. W. Biggar. "Simple Field Methods for Estimating Soil Hydraulic Conductivity," Soil Sci. Soc. Amer. J. 44:3-7 (1980).

16. Biggar, J. W., and D. R. Nielsen. "Miscible Displacement and Leaching Phenomena," Agron. Monograph 11:254-274 (1967).

17. Jury, W. A., and G. Sposito. "Field Calibration and Validation of Solute Transport Models for the Unsaturated Zone," Soil Sci. Soc. Am. J. 49:1331-1341 (1985).

18. Biggar, J. W., and D. R. Nielsen. "Spatial Variability of the Leaching Characteristics of a Field Soil," Water Resour. Res. 12:78-84 (1976).

19. Glotfelty, D. D. "Atmospheric Dispersion of Pesticides from Treated Fields," Ph.D. Thesis, University of Maryland (1981).

20. Spencer, W. F., and M. M. Cliath. "Measurement of Pesticide Vapor Pressures," Residue Rev. 85:57-71 (1983).

21. Bowman, B. T., and W. W. Sans. "The Aqueous Solubility of 27 Insecticides and Related Compounds," J. Env. Sci. Health B14 (6):221-227 (1979).

22. Boynton, W. B., and W. H. Brattain. "Interdiffusion of Gases and Vapors," Int. Crit. Tables 5:62-63 (1929).

23. Bruins, R. "Coefficients of Diffusion in Liquids," Int. Crit. Tables 5:63-72 (1929).

24. Karickhoff, S. W., and D. S. Brown. "Determination of Octanol/Water Distribution Coefficients, Water Solubilities and Sediment. Water Partition Coefficients for Hydrophobic Organic Pollutants," EPA-600/4-79-032 (U.S. EPA, 1979).

25. Gardner, W. H. "Water Content," in Methods of Soil Analysis, C. A. Black, Ed. Monograph 9 (Madison, WI: American Society of Agronomy, 1965).

26. Doorenbos, J., and W. O. Pruitt. "Crop Water Requirements," Irrigation and Drainage Paper 24 (Rome: FAO, 1976).

27. Jury, W. A., W. F. Spencer, and W. J. Farmer. "Behavior Assessment Model for Trace Organics in Soil. I. Model Description," J. Environ. Qual. 12:558-564 (1983).

28. Hamaker, J. W., and J. M. Thompson. "Adsorption," in Organic Chemicals in the Soil Environment, C. A. I. Goring and J. W. Hamaker, Eds. (New York: Marcel-Dekker, Inc., 1972).

29. Kenaga, E. E. "Predicted Bioconcentration Factors and Soil Sorption Coefficients of Pesticides and Other Chemicals," Ecotox. and Environ. Safety 4:26-38 (1980).

30. Lyman, W. J. "Adsorption Coefficient for Soils and Sediments," in Handbook of Chemical Property Estimation Methods, W. J. Lyman, W. F. Reehl, and D. H. Rosenblatt, Eds. (New York: McGraw-Hill, 1982).

31. Rao, P. S. C., and J. M. Davidson. "Estimation of Pesticide Retention and Transformation Parameters Required in Nonpoint Source Pollution Models," in Environmental Impact of Nonpoint Source Pollution, M. R. Overcash and J. M. Davidson, Eds. (Ann Arbor, MI: Ann Arbor Science Publications, 1980).

32. Thomas, R. G. "Volatilization from Water," in Handbook of Chemical Property Estimation Methods," W. J. Lyman, W. H. Reehl, and D. H. Rosenblatt, Eds. (New York: McGraw-Hill, 1982).

33. Hamaker, J. W. "Decomposition: Quantitative Aspects," in Organic Chemicals in the Soil Environment, C. A. I. Goring and J. W. Hamaker, Eds. (New York: Marcel-Dekker, Inc., 1972).

34. Nash, R. G. "Dissipation Rate of Pesticides from Soils," in CREAMS, Vol. 3, W. G. Knisel, Ed. (Washington, DC: U.S. Department of Agriculture, 1980).

35. El Abd, H. "Spatial Variability of the Pesticide Distribution Coefficient," Ph.D. Thesis, University of California, Riverside (1984).

36. Everett, L. G., L. G. Wilson, and L. G. McMillion. "Vadose Zone Monitoring Concepts at Hazardous Waste Sites," Groundwater 20:312-324 (1982).

37. Everett, L. G., L. G. Wilson, and E. W. Hoylman. "Vadose Zone Monitoring for Hazardous Waste Sites," EPA-600/X-83-064 (Las Vegas, NV: U.S. EPA, 1983).

38. Ford, P. J., P. J. Turina, and D. E. Seely. "Characterization of Hazardous Waste Sites - A Methods Manual, Volume II," 2nd edition, EPA-600/4-83-040 (Las Vegas, NV: U.S. EPA, 1983).

39. Howard, P. H., J. Saxena, P. R. Durkin, and L. T. Ou. "Review and Evaluation of Available Techniques for Determining Persistence and Routes of Degradation of Chemical Substances in the Environment," EPA-560/5-75-006 (U.S. EPA, 1975).

40. Mill, T., W. R. Mabey, D. C. Bombwerger, T. W. Chou, D. C. Hendry, and J. H. Smith. "Laboratory Protocols for Evaluating the Fate of Organic Chemicals in Air and Water," EPA-600/3-82-022 (Washington, DC: U.S. EPA, 1982).

41. Plimmer, J. R. "Degradation Methodology: Chemical-Physical Effects," in Proceedings of the Workshop on Microbial Degradation of Pollutants in Marine Environments, A. W. Bourquin and P. H. Pritchard, Eds., EPA-600/9-79-012 (1978).

42. Sherma, J. "Manual of Analytical Quality Control for Pesticides and Related Compounds," EPA-600/2-81-059 (Research Triangle Park, NC: U.S. EPA, 1981).

43. Wilson, J., and M. J. Noonan. "Microbial Activity in Model Aquifier Systems," in Groundwater Pollution Microbiology, G. Britton and C. P. Gerba, Eds., (New York: Wiley-Interscience, 1984).

44. Atlas, R. M., and R. Bartha. Microbial Ecology (Addison-Wesley, 1981).

45. Board, R. G., and D. W. Lovelock. Sampling-Microbiological Monitoring of Environments (New York: Academic Press, 1973).

46. Gilmore, A. E. "A Soil Sampling Tube for Soil Microbiology," Soil Sci. 87:95-99 (1959).

47. McNabb, J. F., and G. E. Mallard. "Microbiological Sampling in the Assessment of Groundwater Pollution," in Groundwater Pollution Microbiology, G. Britton and C. P. Gerba, Eds. (New York: Wiley-Interscience, 1984).

48. Wilson, L. G. "Monitoring in the Vadose Zone. A Review of Technical Elements and Methods," EPA-600/7-80-134 (Las Vegas, NV: U.S. EPA, 1980).

49. Daley, R. J. "Direct Epiifluorescence Enumeration of Native Aquatic Bacteria: Uses, Limitations, and Comparative Accuracy," in Native Aquatic Bacteria: Enumeration, Activity, and Ecology, J. W. Costerton and R. R. Colwell, Eds. (ASTM/STP 695, 1979).

50. Eiland, F. "An Improved Method for Determination of Adenosine Triphosphate (ATP) in Soil," Soil Biol. Biochem. 11:31-35 (1979).

51. Eiland, F., and B. S. Nielsen. "Influence of Cation Content on Adenosine Triphosphate Determinations in Soil," Microb. Ecol. 5:129-137 (1979).

52. Harris, J. C. "Rate of Hydrolysis," in Handbook of Chemical Property Estimation Methods: Environmental Behavior of Organic Compounds, W. J. Lyman, W. F. Reehl, and D. H. Rosenblatt, Eds. (New York: McGraw-Hill, 1982).

53. Jenkinson, D. S., and J. N. Ladd. in Soil Biochemistry, E. A. Paul and J. N. Ladd, Eds. (New York: Marcel-Dekker, Inc., 1981).

54. Leach, F. R. "Biochemical Indicators of Groundwater Pollution," Groundwater Pollution Microbiology, G. Britton and C. P. Gerba, Eds. (New York: Wiley-Interscience, 1984).

55. Stevenson, L. H., T. H. Chrzanowski, and C. W. Erkenbrecher. "The Adeosine Triphosphate Assay: Conceptions and Misconceptions," ASTM/STP 695:99-111 (1979).

56. Trolldenier, G. "The Use of Epifluorescence Microscopy for Counting Soil Microorganisms," in Modern Methods in the Study of Microbial Ecology, R. Rosswall, Ed. (Bulletin from the Ecological Research Committee, Swedish Natural Science Research Council, Stockholm 17:55-59, 1973).

57. Webster, J. J., G. J. Hampton, J. T. Wilson, W. C. Chiorse, and F. R. Leach. "Determination of Microbial Cell Numbers in Subsurface Samples," Groundwater 23:17-25 (1985).

58. Scow, K. M. "Rate of Biodegradation," in Handbook of Chemical Property Estimation Methods: Environmental Behavior of Organic Compounds, W. J. Lyman, W. F. Reehl, and D. H. Rosenblatt, Eds. (New York: McGraw-Hill, 1982).

59. Burkhard, N., and J. A. Guth. "Photolysis of Organophosphorus Insecticides on Soil Surfaces," Pestic. Sci. 10:313-319 (1979).

60. Perdue, E. M. "Association of Organic Pollutants with Humic Substances: Partitioning Equilibria and Hydrolysis Effects in Aquatic and Terrestrial Humic Materials," in R. F. Christman and E. I. Gjessing, Eds. (Ann Arbor MI: Ann Arbor Science Publications, 1983).

61. Perdue, E. M., and N. L. Wolfe. "Prediction of Buffer Catalysis in Field and Laboratory Studies of Pollutant Hydrolysis Reactions," Environ. Sci. Technol. 17:635-642 (1983).

62. Wolfe, N. L., R. G. Zepp, J. A. Gordon, and G. L. Baughman. "Kinetics of Chemical Degradation of Malathion in Water," Environ. Sci. Technol. 11:88-93 (1977).

63. Burkhard, N., and J. A. Guth. "Chemical Hydrolysis of 2-chloro-4, 6-bis(alkylamino)-1,3,5-triazine Herbicides and Their Breakdown in Soil Under the Influence of Adsorption," Pestic. Sci. 12:45-52 (1981).

64. Harris, J. C. "Rate of Aqueous Photolysis," in Handbook of Chemical Property Estimation Methods: Environmental Behavior of Organic Compounds, W. J. Lyman, W. F. Reehl, and D. H. Rosenblatt, Eds. (New York: McGraw-Hill, 1982).

65. Hautala, R. R. "Surfactant Effects on Pesticide Photochemistry in Water and Soil," EPA-600/3-78-060 (U.S. EPA, 1978).

66. Parochetti, J. V., and G. W. Dec, Jr. "Photodecomposition of Eleven Dinitroaniline Herbicides," Weed Sci. 26:153-156 (1978).

67. Smith, C. A., Y. Iwata, and F. A. Gunther. "Conversion and Disappearance of Methidathion on Thin Layers of Dry Soil," J. Agric. Fd. Chem. 26:959-962 (1978).

CHAPTER 3

GENERIC STEPS IN THE FIELD VALIDATION OF VADOSE ZONE FATE AND TRANSPORT MODELS

S. C. Hern
U. S. Environmental Protection Agency, Las Vegas, Nevada

S. M. Melancon and J. E. Pollard
University of Nevada: Las Vegas, Las Vegas, Nevada

INTRODUCTION

The primary emphasis of this document is on the transport and fate of organic chemicals in the vadose zone, i.e., from the soil surface to the groundwater table. Model validation is defined in this report as comparison of model results with numerical environmental data collected in the field or in laboratory observations. Complete model validation requires testing over the full range of conditions for which predictions are intended. At a minimum, this requires a series of validations in various climates and soil types with chemicals that typify the major fate and transport processes. In this chapter, suggested generic approaches to model validation will be presented, but the reader should be aware that many validation problems are specific to a particular site, compound, or model and must be dealt with on a case-by-case basis. In addition, model development and subsequent validation is an evolutionary process by its very nature. Future research into chemical fate and transport processes will produce a refinement of fundamental understanding which will ultimately be used in the update of existing vadose zone models. Methodologies for measurement of model input and output parameters will also be changed and will thus impact model updates and improvement.

The major steps which will be discussed for designing a model validation test are presented in Table 3-1. The reader should remain aware that these generic steps will not always be completed in the sequence that they are here presented. Rather, the validation process may involve dynamic overlap between steps, depending on the data requirements of the model and the selected validation scenario.

Table 3-1. Steps in Field Validation of Soil Fate and Transport Models.

Step 1. Identify Model User's Need--The first step in field validation is to obtain a clear understanding of the model user's need, i.e., how will the model be used.

Step 2. Examine the Model--

Step 2a. Detailed examination of the model: The user must precisely define model input data requirements, output predictions, and model assumptions.

Step 2b. Collect Preliminary Data and Performance of Sensitivity Analysis: Preliminary data are required to conduct a sensitivity analysis and determine the most important input variables.

Step 3. Evaluate the Feasibility of Field Validation--Some models cannot be validated in the field, and the validator should consider this possibility.

Step 4. Develop Acceptance Criteria for Validations--The model user must provide criteria against which the model is to be judged.

Step 5. Determine Field Validation Scenario--Many different approaches to field validation are possible. A scenario should be identified and approved by the model user.

Step 6. Plan and Conduct Field Validations Which Should Include the Following Steps--

Step 6a. Select a Site and Compound(s): Consideration of model input requirements, analytical methods, sources of contamination, and site soil characteristics, etc. are among the many factors to consider in selecting a site and compound(s).

Step 6b. Develop a Field Study Design: Development of a detailed field sampling plan for the specific model compound and site.

Step 6c. Conduct Field Study: Implementation of the field plan is not addressed in these guidelines.

Step 6d. Sample Analysis and Quality Assurance: Many analytical procedures are available depending on the chemical and the matrix. Standardized methods should be used together with a sound quality assurance program.

Step 6e. Compare Model Performance with Acceptance Criteria: A comparison must be made between the performance of the model and the user's acceptance criteria using either graphical or statistical techniques.

STEP 1: IDENTIFY MODEL USER'S NEED

The first step in model validation is to obtain a detailed understanding of the environmental problem confronting a potential model user. This understanding should be acquired through direct discussion with the user. The model validator should elicit from the user the nature of the proposed simulation, how the simulation results will be used, what model input data will be available, how such data will be acquired, and what are the expected model outputs [1].

This problem-identification step is important in defining whether a given model is appropriate, i.e., whether it is capable of meeting the user's needs. Thus, it is essential to have a thorough _a priori_ understanding of the problem to make an assessment regarding the utility of the model. If the model is to be used in several ways or for several different purposes, each use or purpose needs to be defined at the outset. Just because a model has been found valid for one use does not mean it is appropriate for some other use. The validator may discover after a detailed review of the model that the model cannot be used as proposed because of the assumptions of the model the cost of obtaining the input data, etc., and thus may save the user considerable time and expense involved in attempting a field validation.

STEP 2: EXAMINE THE MODEL

This examination should concentrate on determining the required inputs to the model (chemical, biological, and physical), the output predictions made by the model, and the assumptions used in the model construction. The collection of preliminary data, either field measured or obtained from the literature for use in initial sensitivity analysis model runs, is also a part of this early model examination stage.

A model input should be determined for each environmental fate process which affects the compound of interest, e.g., oxidation, hydrolysis, etc. In examining model inputs, it is important to define precisely the units of measure of each input. For example, "organic carbon content" may be expressed on a dry or wet weight basis as g/kg, mg/kg, µg/kg, or as percent data by weight or volume. Other considerations include determining whether the input values of a model should represent a spatial average for some compartment, an isolated point in space, or a soil horizon cross-section, etc., and whether these numerical values represent one point in time or an average over an extended period. Model outputs must be carefully defined by the same criteria as the input data. For example, prediction of x ppm in soil could refer to the total concentration in the soil and soil water matrix or just the soil portion of the matrix.

In addition to examination of model input and output parameters, the relative sensitivity of the model to various input values must be evaluated. Sensitivity is the rate of change of the output of the model caused by a change in input. If a change in an input causes a large change in the output, the model is sensitive to that input; if a change in an input causes a small change in the output, the model is less sensitive to that input. A sensitivity analysis identifies which inputs have a large influence on output and so should be defined with better input accuracy and precision. Using model sensitivities to plan the sampling design and to choose the number of samples can aid in obtaining the appropriate accuracy and precision for the model users' needs.

Sensitivity analysis can range in complexity from a complete factorial design using all combinations of different levels of input parameter values to a simple high and low screening approach such as that described by Hoffman and Gardner [2]. The level of complexity in a sensitivity analysis will depend on the needs of the model user, the model being tested, and the number and type of input parameters required by the model. For a complete factorial-design sensitivity analysis to be practical, the model must have the capability of changing input parameters without extensive user interaction. A simple screening approach to model sensitivity may be the only practical means of optimizing the experimental design of a validation when the model requires user input to change input parameter values for each model run.

The type of input parameters required by the model may also have a profound influence on the extent of a sensitivity analysis. For example, it is possible for a given input parameter to be hypersensitive under one set of environmental conditions and insensitive under another set of conditions. Thus, for parameters of this type, it is necessary to have very good estimates of the range of environmental conditions to be encountered during a validation attempt.

Another major step in the model examination process is development of a thorough understanding of the various assumptions upon which each model is based. Assumptions are commonly made to mathematically simplify models which will affect output interpretations and conclusions drawn from the model simulation results. The major assumptions used in vadose zone fate and transport models are discussed in detail in Chapter 1 of this document. In general, however, it is important for the user at this stage to realize that violation of many assumptions are unavoidable in field validation attempts, especially in the more uncontrolled scenarios. Whether or not violation of these assumptions would result in an invalid test of the model will depend on the user's choice of validation scenario, the sensitivity of model input or output data to the violated assumption(s), and the user's *a priori* established validation acceptance criteria for the model.

STEP 3: EVALUATE THE FEASIBILITY OF FIELD VALIDATION

Field validation is probably the most credible test of a model.

However, inappropriate application of field-tested results can have a significant impact upon the use and credibility of a model, and, in some cases, selection of an alternative model may be the best course of action. The following are examples of instances where the applicability of field validation may be questioned.

- A model that assumes steady state or a dynamic equilibrium may be difficult to field validate fully. Steady state conditions typically do not exist for long in the environment; temperature, infiltration rates, pollutant loads, microbial populations, etc. are almost always changing.

- In some models, input parameters may not be quantifiable. An example of such an input is the active pollutant degrading fraction of the total microbial population. No standard or routine procedure exists for directly acquiring this data.

- Model output data may be uncollectable for purposes of comparison with simulation results. For example, a model may predict the concentration of volatile organic pollutants on clay. Currently, however, no field methods exist for separating clay from sand and silt without affecting the concentration of volatile organic compounds.

- It may be unproductive to field test a model due to a large input sampling error. For example, a degradation rate constant may be measured several times yielding results that vary by orders of magnitude. If this input is sensitive for the chemical and model of interest (a small change in the input results in a large change in the output), the resultant prediction will also have a very large sampling error. This error may exceed the user's acceptance criteria established a priori for precision. However, the output extremes may also reflect the range of conditions which the user could expect to observe in field conditions and may not invalidate the test results.

In summary, it is important to realize that field validation is not always practical and that caution must be used in drawing conclusions from some validation attempts. However, this impact must be weighed against the risks of relying on models which have not been field validated and which may generate output data that are used either with a false sense of security or with undue caution. Persons involved in possible field tests must make this determination on a case-by-case basis.

STEP 4: DEVELOP ACCEPTANCE CRITERIA FOR VALIDATIONS

Prior to attempting to validate a model, the user should develop and provide to the validator those criteria which will be used to accept or reject the mathematical model. After identifying the environmental problem, determining how the mathematical model may assist in resolving that problem, and conducting sensitivity analyses on the model input and output parameters, the user should have a good idea as to the accuracy and precision required of the model. In developing acceptance criteria, the user should consider the predictive ability of alternate physical methods available to solve the problem, e.g., physical construction of a miniature ecosystem by which the environmental problem can be modeled in the laboratory.

The acceptance criteria for a validation should be given in terms of required accuracy, precision, and confidence interval. For example, a certain user may want to estimate the level of some pollutant in the soil within ± one order of magnitude and be correct 95 percent of the time. On the other hand, this same user may set criteria for estimating the migration of a pollutant into ground water at ± a factor of two with the same confidence level, if chemical concentrations in ground water are of greater concern in the scenario of interest. Determination of these acceptance criteria, e.g., confidence intervals about the predictions of the model must also include consideration of the accuracy and confidence intervals associated with the field data to be simulated.

A lack of defined acceptance criteria could result in attempts to validate a model that could not possibly satisfy the user. It may be appropriate to reject a model prior to any field testing after considering the effect of input sampling errors on the predictions of the model. For example, sensitivity analyses as discussed in Step 2 might indicate that the input rate constants would have to be determined an infeasible number of times to achieve satisfactory confidence in the output estimates.

STEP 5: DETERMINE FIELD VALIDATION SCENARIO

There are many possible field validation scenarios that could be proposed. Below, three possible scenarios with their various attributes and limitations are discussed: (1) laboratory testing supplemented with field validations, (2) field validation under controlled conditions, and (3) field validation under natural conditions.

Laboratory Testing Supplemented with Field Validations

This approach involves extensive testing of a model and each of its processes in controlled environments followed by one or a few field tests. Compounds would be selected which undergo several environmental processes. The controlled environment to be used may be a fairly simple laboratory microcosm such as soil columns which offers a relatively high degree of environmental control with varying degrees of complexity and "real world" simulation.

Laboratory testing in controlled environments would also permit detailed study of processes at relatively low expense, and such testing can avoid some problems associated with model assumptions.

The limitations of the scenario are that the results of tests in artificial environments lack credibility since artificial environments only approximate the "real world." A model may perform well under such restricted conditions yet perform poorly in the field since field environments include many variables (e.g., multidimensional chemical flow, soil-type discontinuities, decaying root holes, etc.) that models do not consider in their predictions. The addition of limited followup field testing will add credibility to the laboratory testing. However, this approach may not completely test all processes in the model.

Field Validation: Controlled Conditions

Selection of the controlled field validation scenario would entail dosing natural or seminatural soil systems with various compounds to test each chemical and environmental fate process. In these situations, selected variables such as the loading of test chemicals to the soil system or the delivery of water to small plots or in situ columns (i.e., those planted in external sites) are controlled. Each process in the model is then considered as a submodel and is tested separately. It is possible for results from a test of one process to exceed established criteria confidence levels while all other processes are acceptably accurate.

For any given combination of major soil textural class (sandy loam, clay, etc.), climate, and chemical process, a validation attempt should be conducted. A soil plot could be concurrently dosed with low levels of relatively non-toxic compounds each of which typify one environmental process (e.g., oxidation, hydrolysis, etc.). In such a case, all the model processes relevant to such a test environment could be simultaneously tested. Alternately, the plot and a limited number of chemical processes could be tested. Similar studies would then have to be conducted on other soil types and climates. A few such studies could yield very credible tests at a relatively economical cost compared with those studies based on uncontrolled field validations. However, finding affordable, non-toxic chemicals which represent a simple environmental process in the field is difficult. Furthermore, controlled testing of this type necessitating the addition of contaminants to soil systems may in some cases be scientifically justifiable and environmentally acceptable yet politically impossible. If appropriate test chemicals have been found, however, there are ways to avoid or minimize problems associated with intentional dosing. For example, isolated locations and particularly areas removed from underground drinking water sources might be selected to avoid exposure to the public.

Field Validation: Natural Conditions

Selection of the uncontrolled field validation scenario would entail testing fate and transport models under conditions simultaneously characterized by a combination of soil types, climates, and chemical mixtures and pollutant sources. Such sites would include agricultural fields, pastures, or bare soil lots. These sites might receive contaminant loads from direct dumping of industrial wastes, nonpoint sources (urban and agricultural runoff), atmospheric deposition, or contaminated ground- or surface-water infiltration. The scenario could also include cases where a known type and quantity of organic chemical is distributed across a field, but no controls are placed on climatic or soil conditions. The main advantage of field validation using this scenario is one of credibility. Model validation under "real world" conditions is frequently regarded as the ultimate test of a model. The more complex the environmental conditions and types of chemical mixtures, the more credible is the test. However, such increased complexity generally adds to the cost of testing, increases the sources of errors, makes data interpretation more difficult, and yields less precise predictions.

As in the previous scenario, this approach to model validation would ideally consist of a matrix of validation tests using selected soil types, climates, and compounds to typify each environment and process. A model that satisfies the user's acceptance criteria for each test in the matrix would then be considered valid. This approach, however, would require a large number of field tests to fill in the matrix and thus is very resource intensive.

STEP 6: PLAN AND CONDUCT FIELD VALIDATIONS

This section addresses a number of applied factors that must be considered in the final stages of designing field studies to test a process-type soil fate and transport model. By now, the model has been examined, and all required environmental and chemical inputs identified by process. The outputs of the model are known and all the assumptions and design limitations thoroughly understood. The model has been judged potentially capable of meeting the user's needs, the user's acceptance criteria have been established, and a preliminary decision has been made that field validation is feasible. Finally, a validation scenario has been identified. With these important theoretical preliminaries in mind, the user must consider a variety of specific factors such as selection of field site and chemical compound(s), development and implementation of a field sampling protocol, type of samples to be collected and chemical analyses to be performed, and comparison of model performance with *a priori* acceptance criteria.

Step 6a: Select Site and Compound(s)

Site and compound selection must be considered together. It is most likely that not all the selection criteria can be met at any one location, and the importance of each factor must be weighed by the investigator.

Compound Selection Factors

There are a number of compound-selection factors which are important to consider at this stage, including cost and availability of standard chemical analytical methods, knowledge of the chemical transformations selected compounds undergo in various environmental conditions, compound toxicity, and source of chemical contamination. For example, methods must exist for quantifying input loadings to the model and for determining concentrations of the compounds of interest throughout the soil column or section of vadose zone where field validation is being conducted. In the field validation (controlled conditions) and laboratory testing scenarios, the selection of a pollutant source is obviously no problem since the investigator actually doses the experimental plot. For the field validation under natural conditions, the investigator is restricted to selecting compounds that regardless of source already exist in elevated concentrations and can be tracked through the vadose zone. Precision, accuracy, and compound detection limits within the soil medium need to be determined. Analytical costs can be prohibitive depending on type of desired analyses and required volume of samples.

Availability of reliable rate coefficient information must be considered. In most fate and transport models, rate constants are coupled with site specific environmental parameters to produce site-specific rates. Ideally, development of compound-specific inputs using actual soil and field conditions of the present study should be attempted. If this is not possible, the investigator will have to rely on rate constant information available in the literature. These literature values are typically based on laboratory tests using only the neutral (not ionic) species of the test compound, and, furthermore, different investigator's results often vary over several orders of magnitude. In some cases, rate constant information may not even exist, and an investigator may be forced to select another compound or use rate constants for structurally similar compounds. Excellent sources of rate constant or persistence information are Lyman et al. [3], Rao and Davidson [4], Hamaker [5], and Verschueren [6]. If the latter option is exercised (using values for structurally similar compounds), the user must consider the interpretive limitations of his simulation results.

The environmental fate of the compound is important in all three validation scenarios. Unfortunately, soil chemical processes are highly complex and often inherently linked (e.g., volatilization and water transport), so selection of compounds dominated by a single process is difficult. In all cases, the predicted half-life of the compound must be considered for each process relative to the time during which the compound can be tracked. The movement rate of a compound through a soil matrix is also of primary importance in evaluating environmental fate. A number of physico-chemical factors, e.g., degradation rates, soil-water content, pH, and adsorption coefficients, may affect the half-life or persistence of a compound in the soil. In field validation scenarios, the mode of compound application, watershed management practices, and episodic runoff events near the test area also will influence compound persistence [7].

Compound toxicity and environmental hazard are very important factors when organic chemicals are being applied to natural soil environments. If this approach is pursued and a suitable site located, the least toxic compounds representative of a process should be selected. Since the priority pollutants are undesirable contaminants, other readily degradable pesticides or herbicides that pose no significant threat to the environment should be considered. Compound toxicity can also become a significant problem in laboratory tests in which large quantities of contaminated soil (e.g., from replicated large column tests) are generated. Solid wastes or effluents defined as hazardous by the EPA [8] are subject to extensive federal disposal regulations [8,9] because of their long-term environmental risks. Compound selection criteria may need to include consideration of new disposal restrictions in a large scale laboratory procedure.

Site Selection Factors

Generally, the simpler the site in terms of the amount of data that must be obtained, the more cost effective the validation effort. As previously discussed, one important factor in site selection is that the compound of interest must be present at sufficiently high levels so that it can be traced for a considerable depth or time period. The quantity of pollutant load to the test plot must be known in order to adequately test the model. The accuracy of these data will depend upon the types and number of sources, their relative loading, and their variability. A single uniform surface application is probably the simplest situation while multiple applications perhaps compounded by capillary movement from contaminated ground water or atmospheric deposition create much more complex situations to model. The additional complexity can result in more sources of error, wider confidence intervals around the observed and predicted values, and a less definitive test of the model. Other organic chemicals naturally occurring in the test field may also interfere with chemical analyses of the compound(s) of interest. Soil and interstitial water samples should be analyzed prior to any field study to determine background contaminant levels for the compound(s) of interest. Additional potential sources of contaminants such as irrigation water should be defined and sampled. If interferences are found, a new site must be selected unless clean-up procedures can be developed prior to further sampling efforts and can be maintained throughout the study or unless a way can be found to characterize the level of residues throughout the various sites and depths.

The collection of model site-specific physical or climatic input data, e.g., precipitation or relative humidity, may present no particular problem. However, for many common input parameters, this is not the case [10,11,12,13]. Site access, availability and limitations of appropriate sampling gear, and sample replication limitations dictated by financial or manpower contingencies are other variables of particular consideration. Perhaps the most direct and obvious method of finding a study site is to consult with persons engaged in model development and testing. Other sources of information are the Surveillance and Analysis Divisions of the EPA Regional offices, and State and local waste disposal control personnel. These people may know of sites that meet some or most of the site selection criteria. In locating sites where contaminants can be added to a field, various government-owned reservations, government-sponsored national laboratories, agricultural extension services, private research contractors, and organizations equipped for controlled hazardous waste disposal should be contacted.

Step 6b. Develop a Field Study Design

Type and Number of Samples

The type of samples to be collected is a function of the input requirements of the model, output statements, and the selected test compound and site. Through a detailed examination of the model, an investigator will determine input data requirements by process (oxidation, physical transport, etc.), output predictions, and model assumptions. Chemical tracers, e.g., chloride, fluorescent dyes, etc., may be useful in defining local hydrologic conditions prior to beginning chemical sampling for model validation. The relative number of samples of each type to be collected will be determined by both sensitivity analysis simulation runs which examine parameter sensitivity and variability and by user definition of the desired statistical reliability. The optimum desired number of samples to be collected may have to be tempered by cost and time considerations to give the actual sample size. A detailed description of the statistical rationale underlying decisions about type and number of samples is presented in this document in Chapter 11.

Sampling Equipment

Efficient collection of soil samples in a field study requires a thorough knowledge of the many types of gear available from which to select, characteristics of the soil and field sites to be sampled, and one's ultimate research goals. Technical equipment for unsaturated zone monitoring is frequently not interchangeable from one soil type or environment to another; a few factors influencing equipment selection are summarized for example purposes in Table 3-2.

There have been several excellent documents released in recent years that provide detailed and highly usable information on vadose zone sampler types, limitations, and appropriate application [14,

Table 3-2. Some Factors Influencing Selection of the Most Appropriate Soil Sampling Gear for Vadose Zone Monitoring.

---

1. Soil type - sandy, high clay content, rocky, etc.
2. Soil moisture capabilities - constantly near saturation, cyclic wetting and drying episodes, etc.
3. Site accessibility - roads available for heavy vehicle traffic, permit restrictions, etc.
4. Sample size requirements - number of replicates, grams per replicate needed for analyses, etc.
5. Labor and power requirements - hand-operated, need for electrical generators, weight of sampler when full, etc.
6. Sample depth requirements - surface scrapes only, throughout vadose zone, etc.

15,16,17]. These sources are recommended as invaluable for field studies involving soil monitoring.

Sampling Location

In selecting sampling stations for a field study, rate and the path of movement of a pollutant through the soil compartment to be examined is the first consideration. The investigator needs to determine over what distance and time the compound of interest can be detected. This will be accomplished through a knowledge of the pollutant loading, hydrological conditions, adsorption and chemical breakdown rates, and the detection limit for the compound of interest. Sensitivity analyses, cost, and a knowledge of parameter varability also dictate the number of stations to be included in the monitoring scheme. Availability of appropriate sampling gear (determined by soil type and moisture conditions at the various stations) is an important consideration since, as previously stated, much soil sampling equipment does not have equal utility at different sites or with different soil types. Other sampling criteria influencing site selection include accessibility to the site and labor and power requirements for the preferred soil sampling device.

Duration of Field Project

In designing the length of a field study, one must once again consider the fate and transport characteristics of the chosen chemical(s). Compounds with either short or long half-lives can be used to test physical transport and bioaccumulation processes in the field sampling scenarios. Consideration must be given to the statistical design of the study since determination of the minimum number of field replicates needed for a valid test of the model will influence the duration of the sampling project. The user must decide whether to sample throughout a variety of seasonal and climatic conditions any of which will affect crop growth, hydrology, and type of precipitation (snow vs. rain) or to limit sampling to an abbreviated period of time. Sampling throughout only a single season will simplify the validation process but will also reduce the overall predictive conclusions which can be drawn from validation results.

Step 6c: Conduct Field Study

The actual steps in conducting a field study are highly specific to each individual project. It is impossible to give directions covering all field conditions, so the choice in sampling techniques must often be left to the analyst's judgment. For this reason, detailed instructions for conducting the field study are not discussed here. However, Chapters 4 and 5 of this document present several example scenarios in which the reader can see how various field projects were conducted and resultant data used for model comparisons. Some of the generalized sample collection and logistic details the user should be cognizant of during the actual field study are presented in Table 3-3.

Table 3-3. General Sample Collection and Logistical Considerations for Field Validations.

Sample Collection
- sampling procedures
- sample collection gear - soil cores, augers, etc.
- sample containers - soil bags, moisture cans, etc.
- sample container preparation - soap and water wash, acid rinse, etc.
- sample preservation
- sample coding
- shipment of samples to laboratory
- sample storage prior to analysis

Logistics
- schedules
- record keeping - field notebooks, data forms, etc.
- vehicles - car, trucks, boats, etc.
- maps
- access - permission to sample on private or public lands, road egress, etc.
- notification of local, state, and federal authorities
- personnel - competent team leader, reliable assistants
- data management - how data will be reduced, etc.

It should also be noted that whenever samples are collected at field sites, a variety of parameters important for interpretation and correlation of all data must be recorded in addition to those required for model input. These parameters may include: sampling site location (coordinates, site number), sampling depth, date and time of day, meteorological conditions (air temperature, wind speed and direction, percent cloud cover, precipitation), vegetation type and extent of cover, and surrounding land use. Particular attention should be paid to record-keeping in the field collection portion of the study. All sample containers should be sealed and labeled before they are shipped to the laboratory. Pertinent information should be recorded on a sample tag, e.g., sample number, date and time taken, source of sample, preservation, analyses to be performed, and name of sample collector. The field investigator is also responsible for assuring that adequate size (volume or mass) samples are collected and that the samples are appropriately replicated for meaningful statistical evaluation or quality assurance purposes. Safety procedures (including proper handling protocols and emergency procedures) should be understood and followed by all personnel.

Step 6d: Sample Analysis and Quality Assurance

Chemical Analysis

It is beyond the scope of these guidelines to identify sample collection, sample preservation, and analytical procedures for all organic chemicals that could be used in model evaluations. Detailed analytical procedures have been proposed by the U.S. EPA [18] for determining the concentration of 113 organic toxic pollutants in water. Methods 601-613 apply to the analysis of individual compounds or groups of chemically similar compounds. Methods 624 and 625 are GC/MS procedures for the analyses of the same compounds. The proposed methods cover calibration of instruments, quality control, sample collection, preservation and handling, sample extraction and analyses, and calculations.

Analytical procedures for measuring pesticides in soils are described in detail in Sections 11A and C of a regularly updated EPA pesticide analytical manual [19]; discussion of general procedures from this manual and a number of other relevant sources are summarized in Sections 9A and D of _Manual of Analytical Quality Control for Pesticides and Related Compounds_ by Sherma [20]. New analytical procedures which have been reviewed, tested, and validated by the Association of Official Analytical Chemists (AOAC) are reported in _Official Methods of Analyses_ [21]. A general description of pesticide residue analytical procedures is also provided by Sherma [20] following a discussion on inter- and intra-laboratory quality control. Covered in this document are gas chromatograph procedures and troubleshooting, and procedures for analysis of samples including extraction, isolation, and confirmation of pesticide residues. Although the manual deals primarily with pesticides, many of the procedures and recommendations apply to the analysis of any organic chemical. These are but a few of the many references available for chemical analytical procedures in soils. Where possible, established analytical methods should be used. Whatever method is selected, it must be tested and verified by the analyzing laboratory.

Quality Assurance

The purpose of field sampling in model validation is to provide quantitative environmental input data to the model and to collect field residue data that can be compared to the levels predicted by the model. Such field programs usually include the following operational steps:

- Planning and Conceptualization
- Sample Collection
- Sample Preservation
- Sample Transport
- Sample Storage
- Sample Preparation
- Sample Analysis
- Data Acquisition
- Data Manipulation
- Data Interpretation
- Reporting

To obtain valid data, an overall Quality Assurance (QA) program must be a part of any sampling project. In addition to the usual analytical and equipment QA procedures, a comprehensive QA program should include details on the reliability of the sampling program. Sampling schemes, data analysis strategies, and the objectives of the sampling program must be well-defined for a statistician to assist in the development of an efficient data collection program. A recently published summary QA document for soil studies by Barth and Mason [22] is available; the EPA also has published several quality assurance procedure manuals [23,24,25].

It should be noted that part of the literature data used in the validation process may have been created in the past by different researchers in a variety of studies. While no rigid a priori criteria for acceptance or rejection of these data can be imposed [26], it should be stressed that they must be closely scrutinized. In many instances, data may be of limited value or even useless because the precision and accuracy were not reported or because other key chemical or environmental parameters were not presented.

Step 6e: Compare Model Performance with Field Observations

Comparison of the distribution of organic chemicals present in soils with model predicted values can be accomplished in a variety of ways. The descriptive and statistical methodologies used for comparing model simulations to observed data will be dependent on the characteristics of observed and simulated data sets. For example, rigorous statistical testing for differences between observed and predicted data may not be appropriate for chemical concentration values obtained from the literature due to lack of variance estimates. Similarly, distribution of observed and predicted data sets may be so different that statistical testing of the data yields no more information than a simple graphical presentation would provide (Figure 3-1). If, however, sufficient overlap between observed and predicted data occurs or the shape of the observed and predicted chemical profiles are similar, reasonably simple statistical techniques exist for testing the two data sets for differences or similarities.

Correlation analysis [27] can be used to determine the similarity of the distribution of observed and predicted data. High correlations between observed and simulated distribution curves indicate that the differences among observed data are controlled by factors accounted for by the model [2]. High correlations do not, however, substantiate overlap between observed and predicted data sets. The absolute values of model simulations and observed data may be very different in highly correlated data with differences being a simple scale factor. Replication of observed data at a given depth or time increment allows determination of the significance of the correlation thereby providing additional substantiation for conclusions about model performance.

If sufficient overlap between observed and simulated data sets exists to render difference testing appropriate, a multivariate t-test (Hotellings T^2) can be used [28]. This test provides a means of testing the overall null hypothesis of identical profiles in the populations from which the two groups of data were drawn [29]. Alternatively, confidence intervals about means of observed data can be calculated for a chosen confidence level [27] and plotted with the degree of overlap between observed and predicted data sets being determined by the degree of intersection of plotted confidence bars with the predicted curve.

Unreplicated observed data provide little information for comparison of observed and simulated data. Simple descriptive graphical presentations can be used to display the patterns of similarity or difference between unreplicated data sets, but the significance of this relationship is indeterminate. Similarly, correlations of unreplicated data sets may be high, but the sigificance of this relationship is questionable due to a lack of variance estimates for observed or predicted data.

In many cases, observed chemical profiles in soils may be extremely "noisy" due to spatial and temporal variability. Underlying patterns within the data can be made more obvious by the use of data-smoothing techniques [31]. Since the concentration of a chemical at one depth of soil is often related to the concentration at a subsequent depth, it is possible to remove a portion of the noise or random variation from this type of data using moving averages. This technique may aid in visually displaying the underrlying pattern in a set of observed chemical concentrations in a soil profile which can then be compared graphically with model simulations of that soil profile.

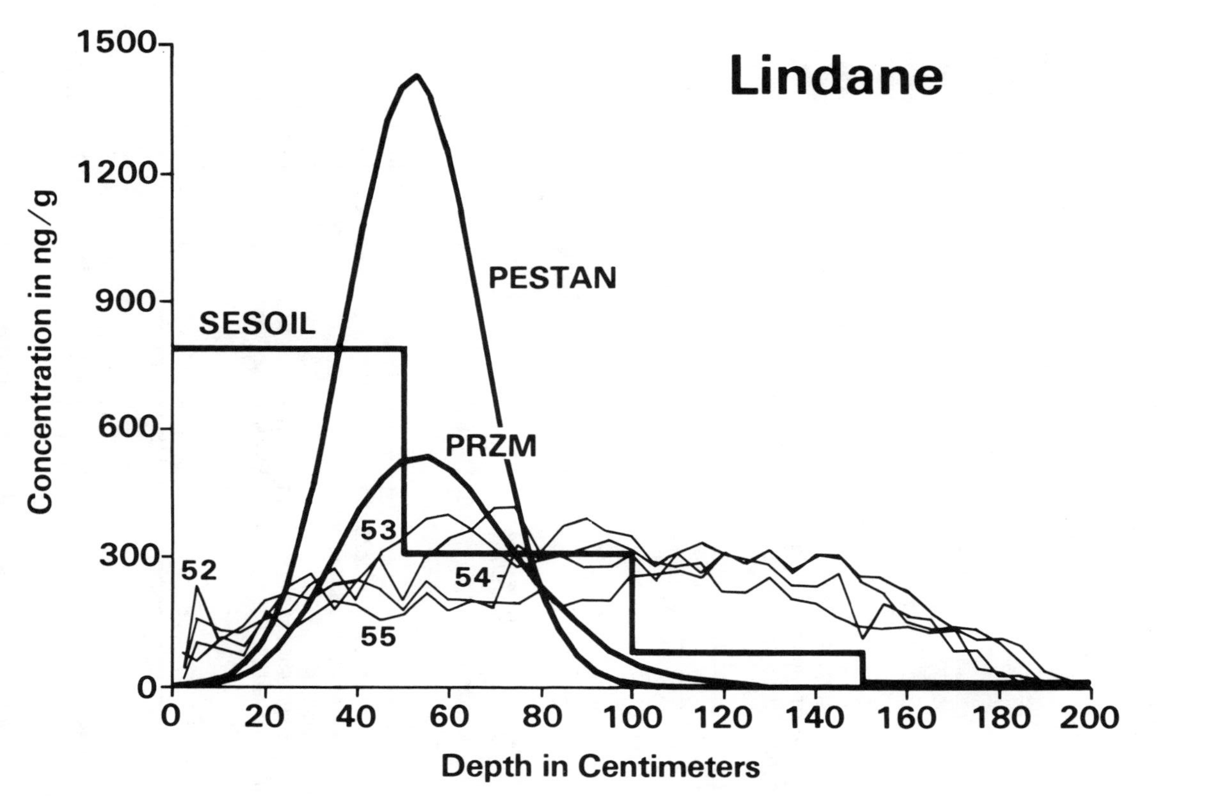

Figure 3-1. Lindane distributed through four soil columns (52-55), compared with SESOIL, PESTAN, and PRZM model predictions (modified from Melancon et al., [30]).

REFERENCES

1. Donigian, A. S. "Recommendation to Improve the Use of Models in Decision Making," in Workshop on Verification of Water Quality Models, T. T. Davis, Ed., EPA-600/9-80-016 (Washington, DC: U.S. EPA, 1980).

2. Hoffman, F. D., and R. H. Gardner. Radiological Assessment. J. E. Till and H. R. Meyer, Eds., NRC No. NUREG/CR-3332, ORNL-5968 (Washington DC: Nuclear Regulatory Commission, 1983).

3. Lyman, W. J., W. F. Reehl, and D. H. Rosenblatt, Eds. Handbook of Chemical Property Estimation Methods: Environmental Behavior of Organic Compounds (New York: McGraw-Hill, 1982).

4. Rao, P. S. C., and J. M. Davidson. "Estimation of Pesticide Retention and Transformation Parameters," in Environmental Impact of Nonpoint Source Pollution, M. R. Overcash, Ed. (Ann Arbor, MI: Ann Arbor Science, 1980).

5. Hamaker, J. W. "Decomposition: Quantitative Aspects." in Organic Chemicals in the Soil Environment," Volume I, C.A.I. Goring and J. W. Hamaker, Eds. (New York: Marcel-Dekker, 1972).

6. Verschueren, K. Handbook of Environmental Data on Organic Chemicals (New York: Van Nostrand Reinhold Co., 1983).

7. Smith, C. N., G. W. Bailey, R. A. Leonard, and G. W. Langdale. "Transport of Agricultural Chemicals from Small Upland Piedmont Watersheds," EPA-600/3-78-056 (Athens, GA: U.S. EPA, 1978).

8. U.S. Environmental Protection Agency. "Title 40 - Protecting Environment. Part 261 - Identification and History of Hazardous Waste" (1983).

9. U.S. Environmental Protection Agency. "Title 40-Protection of Environment. Part 260 - Hazardous Waste Management System: General" (1983).

10. Phillip, J. R. "Field Heterogenicity: Some Basic Issues," Water Resour. Res. 16:433-448 (1980).

11. Parkes, M. E., and J. R. O'Callaghan. "Modeling Soil Water Changes in a Well-Structured, Freely Draining Soil," Water Resour. Res. 16:755-761 (1980).

12. Schmugge, T. J., T. J. Jackson, and H. L. McKim. "Survey of Methods for Soil Moisture Determination," Water Resour. Res. 16:961-979 (1980).

13. Sorooshian, S., and V. K. Gupta. "Automatic Calibration of Conceptual Rainfall-Runoff Models: the Question of Parameter Observability and Uniqueness," Water Resour. Res. 19:260-268 (1983).

14. Wilson, L. G. "Monitoring in the Vadose Zone: A Review of Technical Elements and Methods," EPA-600/7-80-134 (Las Vegas, NV: U.S. EPA, 1980).

15. Everett, L. G., L. G. Wilson, and L. G. McMillion. "Vadose Zone Monitoring Concepts at Hazardous Waste Sites," Groundwater 20:312-324 (1982).

16. Everett, L. G., L. G. Wilson, and E. W. Hoylman. "Vadose Zone Monitoring for Hazardous Waste Sites," EPA-600/X-83-064 (Las Vegas, NV: U.S. EPA, 1983).

17. Ford, P. J., P. J. Turina, and D. E. Seely. "Characterization of Hazardous Waste Sites - A Methods Manual. Volume II," 2nd edition, EPA-600/4-83-040 (Las Vegas, NV: U.S. EPA, 1983).

18. U.S. Environmental Protection Agency. "Guidelines Establishing Test Procedures for the Analysis of Pollutants," Fed. Reg. 44:69464-69675 (1979).

19. Watts, R. R., and J. R. Thompson. "Analysis of Pesticide Residues in Human and Environmental Samples," (Research Triangle Park, NC: U.S. EPA, yearly revisions).

20. Sherma, J. "Manual of Analytical Quality Control for Pesticides and Related Compounds," EPA-600/2-81-059 (Research Triangle Park, NC: U.S. EPA, 1981).

21. Association of Official Analytical Chemists. "Official Methods of Analysis," 14th edition (Arlington, VA: AOAC, 1984).

22. Barth, D. S., and B. J. Mason. "Soil Sampling Quality Assurance User's Guide," EPA-600/4-84-043 (Las Vegas, NV: U.S. EPA, 1984).

23. U.S. Environmental Protection Agency. "Quality Assurance Guidelines for Biological Testing," EPA-600/4-79-043. (Cincinatti, OH: U.S. EPA, 1978).

24. U.S. Environmental Protection Agency. "Interim Guidelines and Specifications for Preparing Quality Assurance Project Plans," QAMS-005/80 (Washington, DC: U.S. EPA, 1980).

25. U.S. Environmental Protection Agency. "The Quality Assurance Bibliography," EPA-600/4-80-009 (Washington, DC: U.S. EPA, 1980).

26. Davis, T. T., Ed. "State-of-the-Art Report of the Hazardous Substances Committee," in Workshop on Verification of Water Quality Models, EPA-600/9-80-016 (Washington, DC: U.S. EPA, 1980).

27. Sokal, R. R., and J. Rohlf. Biometry (San Francisco, CA: W. H. Freeman Press, 1981).

28. Morrison, D. F. Multivariate Statistical Methods (New York: McGraw-Hill, 1976).

29. Harris, R. J. A Primer of Multivariate Statistics (New York: Academic Press, Inc., 1975).

30. Melancon, S. M., J. E. Pollard, and S. C. Hern. "Evaluation of SESOIL, PRZM, and PESTAN in a Laboratory Column Leaching Experiment," Environ. Toxicol. Chem. 5:(In Press, 1986).

31. Davis, J. C. Statistics and Data Analysis in Geology (New York: John & Wiley Sons, Inc., 1973).

CHAPTER 4

EXAMPLE FIELD TESTING OF SOIL FATE AND TRANSPORT MODEL, PRZM, DOUGHERTY PLAIN, GEORGIA[1]

K. F. Hedden
U.S. Environmental Protection Agency, Las Vegas, NV

INTRODUCTION

The previous chapters have described the processes that are important to the fate and transport of organic chemicals in the vadose zone, given an overview of models and modeling and suggested steps to follow in model validation and testing. Chapters 4 and 5 will attempt to show by use of actual examples how some attempts at model validation and testing have been done and will hopefully alert the reader to some limitations and problems which he or she may encounter. A major problem for any validation or testing effort is obtaining a data set that has determined the needed parameters and has sampled the appropriate variables for an adequate period of time. The first example (Chapter 4) describes an effort that was designed with the express purpose of testing a particular vadose zone model, PRZM. In this example, the author will follow insofar as possible the steps in field validation described in Chapter 3 and outlined in Table 3-1. This effort is not yet complete, so only the initial stages can be described, but these should serve to give the reader an idea of what is involved in testing a model. The second example (Chapter 5) describes two retrospective studies. It gives guidance on finding and selecting an adequate data set and illustrates some of the problems encountered when applying retrospective studies to model validation attempts.

---

[1]This example is primarily based on "Pesticide Migration in the Unsaturated and Saturated Soil Zones. Part I. A Field Study to Support Model Development and Testing, Dougherty Plains, GA, 1983" by Sandra C. Cooper, Robert F. Carsel, Charles N. Smith, and Rudolph S. Parrish [1].

STEP 1: IDENTIFY NEEDS

The Dougherty Plain project is an ongoing five-year cooperative research project between the U.S. EPA, Environmental Research Laboratory, Athens, Georgia, and the U.S. Geological Survey (USGS) in Georgia. The project was designed to develop a data base for testing mathematical models that evaluate the potential for contamination of ground-water resources from increased pesticide applications. The model being utilized for initial testing is the Pesticide Root Zone Model (PRZM), developed by Carsel et al. [2]. PRZM simulates the transport of pesticides in the unsaturated zone by integrating hydrologic and pesticide process data relating to the sorption, degradation, and leaching properties of pesticides. The Dougherty Plain data will be used to calibrate, test, and refine PRZM.

STEP 2: EXAMINE MODEL

Input requirements necessitate the collection of field data on the soil chemical and physical properties and the existing ground-water conditions. Certain climatic data are also required. For comparing model predictions with field data, samples collected in the unsaturated and saturated zones after pesticide application are also needed. Model input data requirements are listed in Table 4-1. The PRZM user's manual [2] provides much guidance in obtaining required input data. The EPA Athens Lab also has developed a meteorological information data base that can provide all needed meteorological data. The Soil Conservation Service (SCS) has a soils data base that can provide most of the needed soil cropping information.

Model output predictions are listed in Table 4-2. The hard output may be called by day, month, or year. Three files may also be called: (1) hydrologic, (2) pesticide output, and (3) pesticide concentration. Values are printed by time step and compartment with summary data, and a variety of plots may be produced [2].

Several simplifying assumptions have been made in developing the basic equations of PRZM. While these assumptions may not exactly describe the process modeled, they greatly simplify the model. Sorption of pesticide on soil particles is assumed to be instantaneous and reversible. Dispersion and diffusion are combined and described using Fick's law, and they are assumed constant. One first order decomposition rate is used to represent the sum of all degradative processes in both the soil and water phases. Plant uptake of pesticides is modeled by assuming that uptake of a pesticide by a plant is directly related to the transpiration rate. The equations are written for vertical movement at a single point, but field soils are quite variable, and conditions at one point are not necessarily representative of conditions at other points.

A sensitivity analysis was conducted to determine which input parameters should be given attention, and to what extent variations

Table 4-1. Input Data Requirements for PRZM.

Meteorological Data
- Daily precipitation, cm
- Daily pan evaporation, cm } (both not necessarily needed)
- Daily average temperature, °C }

Hydrology Data
- Pan factor, dimensionless
- Snow factor, cm snowmelt/°C above freezing
- Minimum depth to which evaporation is extracted, cm
- Average daily hours of daylight for each month, hr/day
- Universal soil loss equation parameters/factors, dimensionless
  K-Soil erodability; LS-Topographic/length-slope;
  P-Supporting/Conservation Practice; C-Cover and Management
- Field area, ha
- Average duration of runoff hydrograph from runoff producing storms, hr
- Maximum interception storage for each crop, cm
- Maximum active root depth for each crop, cm
- Maximum areal coverage of each crop at full canopy, (percent)
- Runoff curve number for antecedent soil water condition for fallow, crop, and residue fractions of the growing season, dimensionless
- Maximum dry foliage weight of each crop at full canopy, kg/m^2
- Number and dates of crop emergence, maturation, and harvest for each cropping period, dimensionless

Pesticide Data
- Number of applications, dimensionless
- Date, amount, and depth of each application; dimensionless, kg/ha, cm
- Type of application, dimensionless
- For foliar application: decay rate on foliage, foliar extraction coefficient, and filtration parameter; $days^{-1}$, cm^{-1}, $m^2\ kg^{-1}$
- Pesticide solubility; may be used to estimate partition coefficient by any of 3 methods, mole fraction or mg l^{-1} or μmoles l^{-1}
- Pesticide/soil sorption partition coefficient per horizon, $cm^3\ g^{-1}$
- Pesticide decay rate for each horizon, $days^{-1}$

Soil Data
- Depth of core, cm
- Plant uptake efficiency, dimensionless
- Total number of soil horizons and compartments
- Thickness, bulk density, hydrodynamic dispersion, initial soil water content, and drainage parameter per horizon; cm, g cm^{-3}, $cm^2\ day^{-1}$, $cm^3\ cm^{-3}$, day^{-1}
- Field capacity, wilting point, organic carbon content, percent sand and clay in each soil horizon; $cm^3\ cm^{-3}$, $cm^3\ cm^{-3}$, percent, percent, percent
- Initial pesticide level in each compartment, mg kg^{-1} or kg ha^{-1}

Table 4-2. The Major Output Prediction Functions Currently Performed by PRZM.

1. Calculation of soil-water characteristics based on textural properties
2. Calculations of K_D based on water solubility models
3. Echo of inputs to output files
4. Determination of crop root growth
5. Specific time series data output
6. Crop interception of rainfall
7. Division of precipitation between rain and snow
8. Calculation of evapotranspiration
9. Snowmelt computation
10. Calculation of plant uptake factors
11. Determination of curve number from cropping period and soil moisture
12. Computation of runoff and infiltration
13. Calculation of soil hydraulics
14. Calculation of pesticide transport in soil
15. Total pesticide application
16. Water and pesticide mass balance computation
17. Output of fluxes, storages, etc
18. Input checking
19. Foliar pesticide application decay and washoff
20. Soil erosion and erosion pesticide loss

in expected input values would affect output values. The sensitivity analyses conducted at this stage of the model examination were based on the range of possible values for each input parameter. Table 4-3 lists some of the more sensitive parameters identified during this analysis stage. Further sensitivity analyses were then done using a range of probable values for more sensitive parameters which were either taken from the literature or measured in analysis of preliminary samples. The input parameters most sensitive with regard to leaching are the decay rate, sorption coefficient, and depth of active layer (root zone).

STEP 3: FIELD TESTING FEASIBILITY

PRZM was designed as a site specific model using field data [2] The user's manual and the model itself also lend themselves to determination of needed parameters by several alternate methods. For example, field capacity, wilting point, and saturation soil-water contents may be calculated from soil textural information. Also, the sorption coefficient, K_D, may be calculated from a chemical's aqueous solubility by any of three different methods. The output variables that need to be measured for comparison with model output are the soil-water content, pesticide content in the soil column, and pesticide concentration in the soil water. These variables can

Table 4-3. Sensitive PRZM Input Parameters.

Category	Parameter
Transport	KD (Sorption Coefficient)
	BD (Bulk Density)
	THEFC (Field Capacity)
	THEWP (Wilting Point)
	CN (Curve Number)
Supply	RA (Application Rate)
	KS (Decay Rate)
	AL (Active Layer-Root Zone)

be measured from soil cores and soil-water samples collected with suction lysimeters [3]. The amount of pesticide leaching past the root zone, which may be measured using the above procedures, is that amount of pesticide which may potentially contaminate ground water.

STEP 4: CRITERIA OF ACCEPTANCE

The participants of the Predictive Exposure Assessment Workshop sponsored by the U.S. EPA in Atlanta, Georgia, on April 27-29, 1982, concluded that for screening applications, a model should be able to replicate field data (concentration profile, total mass, flux past root zone, soil-water content and storage, etc.) within an order of magnitude and site specific applications within a factor of 2. The Dougherty Plain project staff elected to use similar criteria. Therefore, if the model simulates the field data outside more than a factor of 2, the model will need to be modified to the extent needed to meet this criterion.

STEP 5: DETERMINE FIELD VALIDATION SCENARIO

The example study is a field validation under natural conditions in the Dougherty Plains area of southwest Georgia. PRZM has been tested a number of times previously (see Donigian and Rao, Chapter 5); however, most of these studies utilized data that had been collected for purposes other than model testing. It was decided that a field validation under natural conditions would offer maximum credibility and acceptance in the modeling community.

STEP 6a: PLAN AND CONDUCT FIELD TEST

Select Study Area and Compound(s)

The following criteria were used in the selection of a representative field site for evaluation of PRZM. The study area should (1) be in a major agricultural area with potential for leaching to ground water; (2) have relatively easy access to the Athens lab, (3) should be composed of multilayered soils in order to adequately test the model, (4) should be relatively flat topography in order to minimize runoff, thus simplifying mass balance determinations as well as making it less expensive to monitor, (5) should be a manageable size (less than 8 ha), (6) should have a shallow water table, (7) should be isolated from domestic water supplies for safety reasons, (8) should be located within close proximity to a ground-water divide to assure unidirectional ground-water flow, and (9) should be available for lease for a period of five years in order to obtain an adequate amount of data. An Interagency Agreement was developed between the U.S. EPA in Athens, Georgia, and the USGS District Office in Doraville, Georgia. The USGS has a field office in Albany, Georgia, which had just completed a five-year irrigation study, so staff were familiar with fields in the area and knew many of the land owners. Using the criteria set up by EPA describing a desirable field site, the USGS identified several potential study areas. These sites were then visited and ranked by EPA. A 3.9 ha field was selected in southeast Lee County (near Albany) in the Dougherty Plain area of Georgia (Figure 4-1), and a five-year lease agreement obtained with the land owner. The topography of the area is gently undulating uplands and plains, and low-lying plains that are either well drained or swampy [4]. The re-relief of the Dougherty Plain is low to gently sloping with maximum relief not exceeding 24 meters and slopes of less than 2 percent. Agriculture accounts for 80 percent of the land use within Dougherty Plain [1].

In order to avoid legal complications, the pesticides selected for use would need to be registered for land application to the selected crop. The initial chemical chosen should be sufficiently mobile to assure that chemical movement will be observed during the study period; if additional chemicals are chosen they should span a range of lesser mobilities.

Two pesticides were chosen according to the previously established criteria. They included the insecticide, aldicarb 2-methyl-2(methylthio) propionaldehyde O-(methyl-carbamoyl)oxine, and the less soluble less mobile herbicide, metolachlor 2-chloro-N-(2-ethyl-6-methylphenyl)-N-(2-methoxy-1-methylethyl)acetamide. Both aldicarb and metolachlor are widely used in the Dougherty Plain area. The selected crop was peanuts (*Arachis hypogaea*).

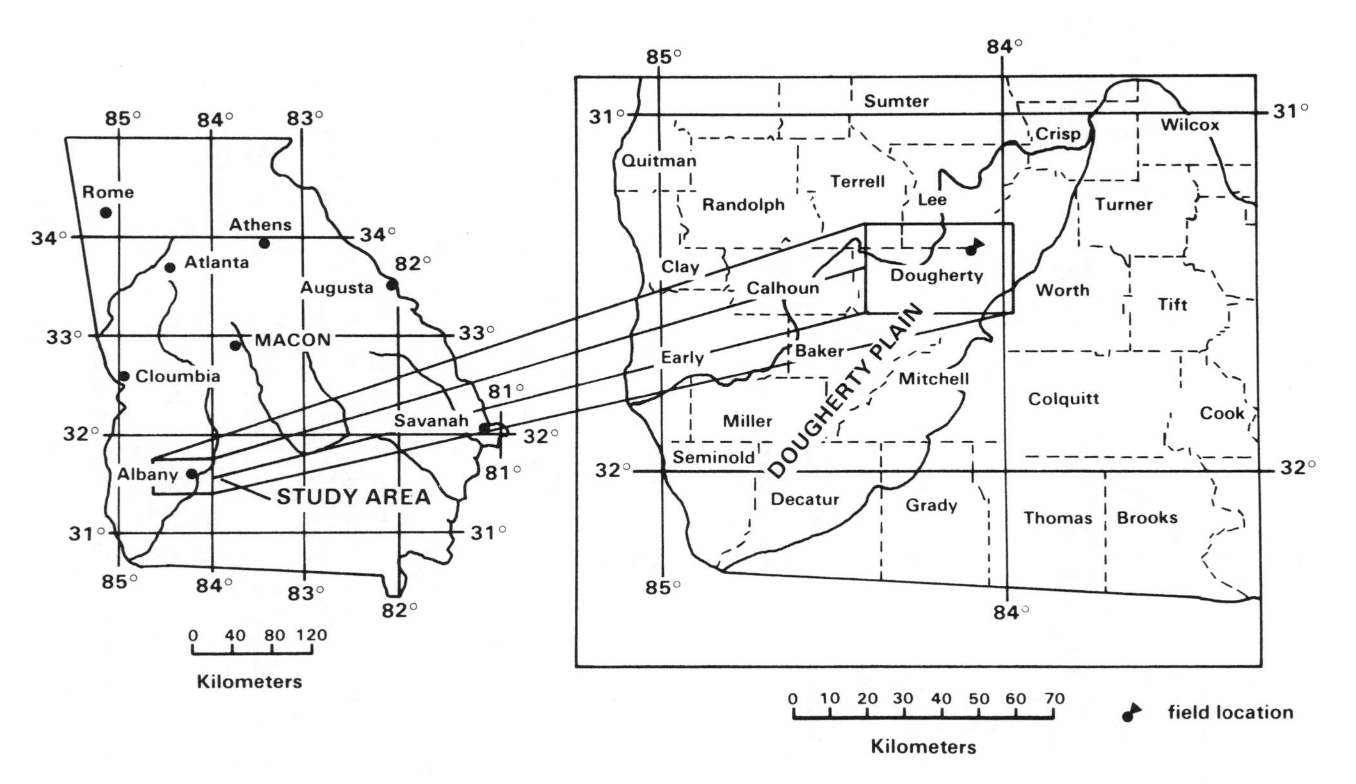

Figure 4-1. Study site location.

Aldicarb

Aldicarb is a systemic carbamate insecticide used in the control of soil insects and nematodes in cotton, peanuts, potatoes, and other crops [5]. The registered use of aldicarb on peanuts is summarized in Table 4-4. The chemical is applied to soil in the form of granules containing 10 percent or 15 percent active ingredient. It is a noncorrosive, non-flammable compound that is stable except in the presence of strong alkali [6]. Aldicarb is metabolized by both oxidative and hydrolytic processes. Oxidation of aldicarb produces toxic derivatives of aldicarb sulfoxide and aldicarb sulfone. Aldicarb sulfoxide and aldicarb sulfone are both acetycholinesterase inhibiting compounds. The term "Total Toxic Residues" (TTR) has been used to refer to aldicarb and its derivative compounds [7]. Aldicarb and the sulfoxide may be oxidized to the sulfone which is then analyzed by HPLC as TTR [1]. Table 4-5 lists some of the chemical and physical properties of aldicarb and its sulfoxide and sulfone derivatives.

A number of laboratory studies have examined the fate of aldicarb residues in soils [8,9,10]. Supak [11] conducted laboratory investigations on the volatilization, degradation, adsorption, and desorption characteristics of aldicarb in soils and clays. Field studies on the fate and persistence of aldicarb and its metabolites in soils have been conducted by a number of other researchers [12, 13,14,15,16,17].

The Union Carbide Corporation has collected laboratory and field data concerning the environmental impact of aldicarb. The U.S. EPA [18] reported that aldicarb is oxidized in the soil to the toxic metabolite aldicarb sulfoxide. The sulfoxide is largely hydrolyzed to nontoxic oximes and nitriles while a small percentage is oxidized to aldicarb sulfone. The sulfone also degrades into nontoxic nitriles.

Metolachlor

Metolachlor is a selective herbicide used for weed control in corn, soybeans, peanuts, and certain other crops [5]. The acute toxicity of metolachlor to mammals is shown in Table 4-6.

Because metolachlor is a relatively new compound, limited information is available. Several studies have shown, however, that it is readily sorbed to clay and organic matter. The Chemistry Branch at Athens developed a GC method for analyzing metolachlor in soil. Metolachlor will be used only in application monitoring.

STEP 6b: DEVELOP A FIELD STUDY DESIGN

The Dougherty Plain project was designed to: (1) monitor leaching, (2) evaluate spatial variability of pesticide application and soil properties, and (3) establish sorption coefficients and

Table 4-4. Registered Agricultural Uses of Aldicarb on Peanuts [18][a].

Pests Controlled	Ounces of Formulation per 1000 ft of row[b]		Pounds of Active Ingredient Per Acre (range)	Application
	10%	15%		
Trips	11-22	7.5-15	1.0-2.1	Apply granules in seed furrow and cover with soil. In south-west use high rate only.
Nematodes (root-knot ring, lesion, string, stunt, spiral and stubby-root)	22-33	15-22	2.0-3.0	Apply granules in a 6- to 12-inch band and work into the soil or cover with soil to a depth of 2 to 4 inches. Plant seeds in the treated zone.

[a]0.05 ppm (peanuts), 0.5 ppm (peanut hulls), 90 day preharvest interval. Use high rate of heavy organic or clay soils. Applied as soil application at planting. Granules should be worked into the soil to a depth of at least 2 inches or covered with soil to a depth of at least 2 inches to provide maximum performance and minimize the hazard to birds. Do not allow livestock to graze in treated areas before harvest. Do not allow hogs to root in treated fields. Do not feed peanut hay or vines to livestock. Do not plant any crop not listed on label in treated soil within 100 days of last application. Do not use in house or home gardens.

[b]Pesticide applied at planting using 36-inch row spacing.

Table 4-5. Chemical and Physical Properties of Aldicarb and its Sulfoxide and Sulfone Derivatives[a].

Property	Aldicarb	Aldicarb Sulfoxide	Aldicarb Sulfone
Melting point (°C)	98-100	108-110	134-136
Vapor pressure (25°C, mm, Hg)	10^{-4}	$10^{-4.15}$	$10^{-4.25}$
Solubility (%, 27°C)			
Water	0.6	25.0	0.7
Acetone	28.0	27.0	5.1
Trichloromethane	38.0	39.0	3.2
Methyl alcohol	NA	56.0	3.0
Oral LD_{50}, rats (mg/kg)	1.0	< 1.0	10.0
Clay reaction	Negative montmoril-lonite	Negative montmoril-lonite	positive koalinite
Biological decay	7-56 days, first order	7-56 days, first order	50-1900 days, first order

[a][1,11]

Table 4-6. Toxicological Properties of Metolachlor[a].

	Metolachlor Technical	Dual 8E
Acute oral LD_{50} (Rat)	2780 mg/kg	2534 mg/kg
Acute dermal LD_{50} (Rabbit)	>3100 mg/kg	>3000 mg/kg
Acute aerosol inhalation LC_{50} (Rat) (4-hour exposure)	>1.75 mg/l	>0.94 mg/kg

[a][19]

degradation rate constants for the two selected pesticides. The experimental area consisted of a 3.9 ha field with a 9.1 m (30 ft) border around the edge of the field to eliminate potential boundary influence (Figure 4-2).

A soil survey of the field conducted by the local SCS office was used as the basis for selecting the primary monitoring sites and additional soil sampling sites. The total area in each soil series is shown in Table 4-7. The sampling size of 30 has been determined to be a reasonably large number of sites to adequately monitor a field [20]. Due to management practicality and resource limitations, the 20 sites shown in Figure 4-2 were finally selected. The degree of precision lost in going from 30 to 20 sites was within an acceptable range according to Bresler and Green [20]. Subjectively, this number also seemed to provide reasonable coverage of the field especially insofar as preliminary investigation was concerned. Additional information on the Dougherty Plain field site is provided by Carsel et al. [21].

The area of each of the three major soil series identified in the field was divided by the total field area to calculate relative proportions of area for each soil series in Table 4-8. The 20 sampling sites were allocated to the three series according to these proportions. Four additional sites were selected for soil characterization so that a total of 24 sites were used for soil characterization.

A 15.2 meter (50 foot) grid was used to select the array of sampling sites. Grid points that were within 6.1 meters (20 feet) of a soil series boundary were eliminated from consideration. This yielded a total of 96 candidate grid points out of the total of 134 grid points. The number of candidate grid points in the three series is shown in Table 4-8. The samples for each soil series were selected randomly from the candidate points for that soil series. The resulting set of 20 selected sites became the primary sampling sites; proportional allocation of the 20 random sites produced 5, 6, and 9 samples for the Tifton, Ardilla, and Clarendon series, respectively.

Alternate sites were then randomly selected from the remaining grid points. These sites were to be used in the event that one of the primary sites proved to be unacceptable. Permanent reference points were surveyed along the perimeter of the field in the area outside the 9.1 meter border. The reference points were to be used in conjunction with a transit to relocate sampling equipment after planting in the spring of each study year.

Soil core samples for physical characteristics (sand, silt, clay, organic carbon, and pH) were taken at 24 sites at four depths to 1.2 m. The analysis was done by the University of Georgia Soil Test Laboratory. Background soil core samples for aldicarb and metolachlor were taken at the 20 primary sites; 160 samples were taken in 15 cm increments to 1.2 m. Samples were analyzed by the OPP Laboratory at Beltsville, Maryland, for aldicarb and at the

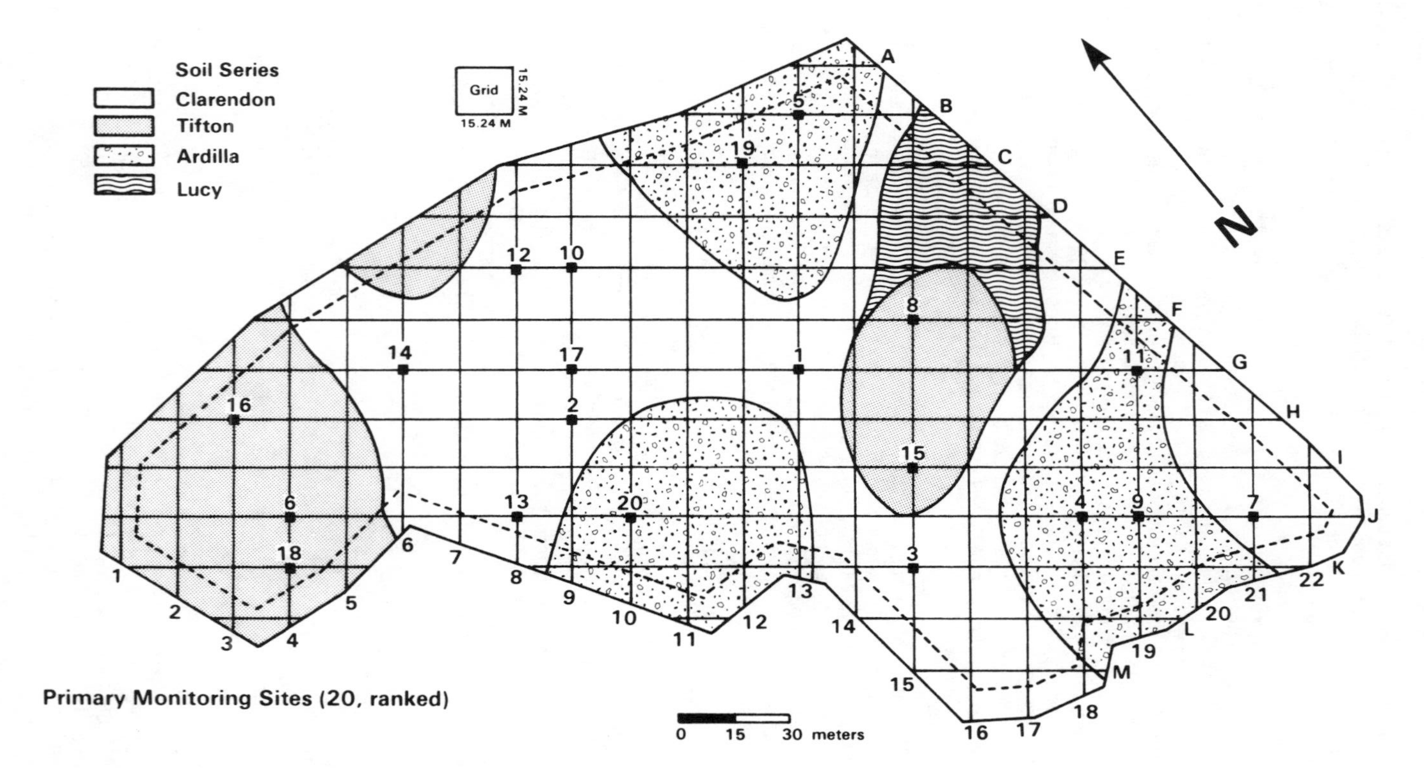

Figure 4-2. Dougherty Plain field stie.

Table 4-7. Area (ha) Per Soil Series in Daugherty Plain, Georgia.

Series	Ardilla	Clarendon	Tifton	Lucy	Total Field
Area within 9.1-meter border	0.90	1.43	0.65	0.14	3.12
Area outside 9.1-meter border	0.22	0.31	0.21	0.04	0.78
Total acreage	1.12	1.74	0.86	0.18	3.90

Environmental Research Laboratory, Athens, for metolachlor. Samples for degradation and sorption studies were taken with well drilling equipment in duplicate at four depths to 1.2 m using the method described by Wilson [3]. The sorption and degradation parameters for aldicarb and metolachlor using these soil samples were measured by Rao et al. [22]. Undisturbed core samples were taken from a minimum of seven sites on the field perimeter representing each soil series for determination of bulk density and soil-water characteristic curves. Pits were excavated to a depth of 1.2 m, and samples were taken from the sidewalls of the pits. Three soil cores and one bulk soil sample were collected from each horizon. The analysis was done by the R.S. Kerr Environmental Research Laboratory in Ada, Oklahoma. Further background water samples were taken (10 in the residium and 4 in the Ocala aquifer). Analysis was done by U.S. EPA in Beltsville and Athens.

Table 4-8. Sample Allocation by Soil Type in Daugherty Plain, Georgia.

Soil Type	Proportion	Number Samples for Primary Sites	Number Candidate Grid Points	Number Samples for Additional Sites
Tifton	0.23	5	20	1
Ardilla	0.30	6	30	1
Clarendon	0.47	9	46	2

Twenty-one permanent wells were installed at 15 of the 20 primary sites selected by priority number. In the shallow saturated zone, these wells were located at depths of 4.6 and 6.2 meters, and four additional wells were located outside the perimeter in the Ocala aquifer at 53.3 meters. These wells were established to characterize the hydrology of the site by providing information on ground-water depth and direction of the flow as well as indicating if any chemical leaches to ground water. The ground water data do not input to PRZM but do further characterize the site. The use of suction lysimeters reduces the time and manpower needed for sample collection; it also helps preserve the integrity of the field. For monitoring by use of suction lysimeters, see Wilson [3]. Figure 4-3 shows a schematic of a monitoring site.

A weather station equipped with a hygrothermograph, an evaporation pan, an anemometer, and two rain gauges was installed on the north side of the study field. This permitted measurement of air temperature, relative humidity, evaporation, wind speed, and precipitation on a continuous basis.

Monitor Leaching

Soil cores were collected at each sampling site to a depth of 1.2 m (in 15-cm increments). Samples were collected from suction lysimeters located at even intervals between the coring depth and the normal water tables at 1.5, 2.1, and 2.7 meters, and at wells located at depths of 4.6 and 6.2 meters.

Spatial Variability

The spatial variation of aldicarb (granular formulation) application at the time of planting was measured by collecting soil samples on the same day along the band where aldicarb was applied. The samples were collected by pressing an aluminum can into the soil. Samples were systematically taken in a three row square grid at each sampling site.

The spatial variation of metolachlor application (emulsifiable concentrate) was measured by placing 200 (18.5 cm diameter) filter discs in duplicate at 100 sites at even intervals in the field prior to application using a systematic sampling plan. The use of filter discs to monitor pesticide application using boom-type spraying equipment has been successfully demonstrated in previous field studies [23]. Filter disc as well as other techniques for pesticide application monitoring are discussed in greater detail by Smith et al. [24]. Metolachlor was limited to application monitoring due to resource constraints.

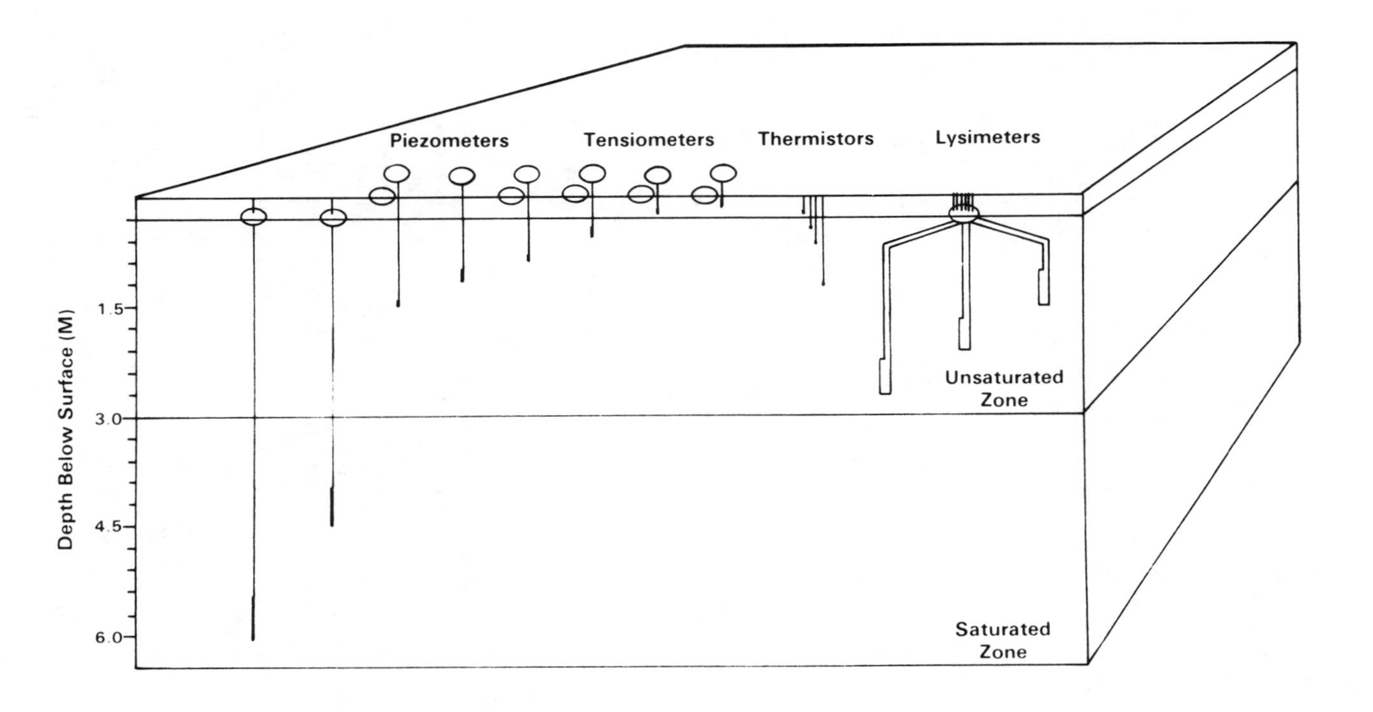

Figure 4-3. Schematics of monitoring site.

Degradation and Sorption

Degradation and sorption studies for both aldicarb and metolachlor were conducted under a cooperative agreement between the U.S. EPA and the University of Florida. Other than the amount of precipitation, these two parameters are the most important in determining if a pesticide will leach. Well-drilling equipment was used to collect duplicate samples at each of the twenty primary sites at four depths. Samples were taken before the start of the main study. Cores were extracted using a Frazier sampler supplied by the Robert S. Kerr Environmental Research Laboratory of the U.S. EPA in Ada, Oklahoma. Using aseptic procedures [3], the first five cm of each soil core was discarded in order to avoid contamination, and the center of the soil core was extracted into sterile Mason jars for each depth sample increment. Degradation rates were determined as a function of depth. The outside parts of the soil core were collected for sorption studies. Soil samples were used to determine the field spatial variation of sorption coefficient, organic carbon content, and pesticide degradation rate coefficients [22].

STEP 6c: CONDUCT FIELD STUDY

In addition to characterization and background work activities in the main field during 1983, two 15.2 by 15.2 meter test plots (called SP1 and SP2) were planted in an adjacent field and treated with aldicarb. The test plots were used to evaluate the effectiveness of the soil solution samplers (lysimeters), to develop soil sampling techniques, and to gather background data on aldicarb movement in soils. These preliminary field activities will tend to modify and fine tune the original field study design.

These two study plots were located in a 8.1 ha field directly north of the main study area. Prior to planting, the field was treated with a mixture of Balan and Vernam 7E. These are pre-emergent herbicides incorporated in the soil for seasonal control of annual grasses and broadleaf weeds. On May 20, 1983, the farm owner planted the field with 1200 pounds of peanuts (Florunner variety) and applied aldicarb at a rate of 0.56 kg ha^{-1} active ingredient. The peanut rows were spaced 0.9 meter apart. Shortly after planting, a mixture of herbicides Alachlor and Dyanap was surface applied to the field. In addition, Bravo, a fungicide, plus elemental sulfur were foliar-applied to the peanut crop approximately one month later.

Soil core samples were collected for aldicarb residue analysis at 6, 20, and 38 days after planting and application of aldicarb. Soil samples were taken within two days after rainfall events occurred so as to allow time for soil to reach field capacity soil-water content. Four soil cores were taken from each plot to a depth of 1.2 meters. Soil samples were taken on two adjacent rows, and additional samples were taken at 30-cm intervals across the 0.9-meter width perpendicular to the rows of peanuts. The same rows were sampled in both plots, SP1 and SP2. A five cm diameter core barrel attached to a Giddings hydraulic soil sampler mounted on the back of a four-wheel drive truck was used to extract the soil cores. Two cores were collected on the row of peanuts where aldicarb was previously applied, and two cores were collected between the rows.

After each soil core was collected, a metal extracting rod was used to push the core out into a stainless steel sample tray. Between samples, the core barrel, the extracting rod, and the sample collection tray were scrubbed and rinsed with acetone. The core was subdivided into 15-cm sample increments and then transferred into aluminum cans and mixed thoroughly. The cans were sealed with plastic tape. Labels on each can listed the sample location, sample depth, and the date. The metal cans then were transferred into plastic bags; the bags were sealed and placed in coolers of ice. All samples were frozen prior to shipment to the analytical laboratory.

Two types of lysimeters were considered for use in monitoring the unsaturated soil zone. These experimental lysimeters were installed in SP1 to test and evaluate their respective effectiveness and efficiency. The lysimeters under consideration included (1) a polyvinyl chloride (PVC) plastic pipe fitted at one end with a porous ceramic cup and with a two-holed rubber stopper (for polyethylene tubing) attached to the opposite end, and (2) an all Teflon® type device including filter. Both lysimeters were found unacceptable for use in this study. The PVC lysimeter sorbed chemical while the Teflon® lysimeter had a very low air-entry value. Therefore, a stainless-steel lysimeter was designed, tested, and then after proving acceptable installed in the main field [25].

During the 1984 growing season, samples were collected for aldicarb and metolachlor consisting of (1) a field "shakedown" process for development of sampling techniques, shipment/storage practices, calibration/equilibration of monitoring equipment, and definition of team responsibilities between project participants; and of (2) quantitatively monitoring the movement of pesticide residue within the soil profile for TTR for aldicarb. Samples were also taken to evaluate the spatial variability of a granular pesticide application technique (aldicarb) and the spatial variability of a surface application technique (metolachlor).

Soil and ground-water samples were collected in February 1984 from the main (3.9 ha) site for background residue analysis. A soil core was taken to a depth of 1.2 meters at each of the 20 monitoring sites. The soil core was subdivided into 15-cm depth increments that were transferred into aluminum cans, sealed, labeled, and placed in coolers of ice.

Ground-water samples were collected from 10 stainless steel monitoring wells, selected by priority number, in the residue and from the four wells open in the Ocala aquifer. The wells were evacuated with a small portable gasoline-powered pump for approximately 15 minutes to remove at least one bore-volume of water from the well. A one-liter volume of sample was collected by an all-Teflon® point-source bailer. The samples were transferred into one-liter Nalgene® bottles, and were then placed in coolers of ice.

After metolachlor application, the filter discs were collected and sealed in Mason jars, placed in ice coolers, and sent to the analytical laboratory for analysis.

All samples collected for degradation or sorption studies were stored in ice chests and shipped to the University of Florida for analysis.

Soil cores were taken during the growing season for aldicarb analysis within two days after major rainfall events occurred.

All soil core and ground-water samples were frozen prior to shipment in order to preserve sample integrity. The frozen samples were shipped in ice to the Beltsville Laboratory for background residue analysis of aldicarb.

Two to three plants were taken for analysis from each sampling site for analysis in order to evaluate pesticide loss by plant uptake. They were frozen and shipped in ice with the soil cores and water samples.

To effectively collect sufficient data that will provide a quantitative statement of both model performance and pesticide leaching in Dougherty Plain soils, a minimum of three more years of the present, ongoing, highly characterized pesticide monitoring should be completed. The continuation of the project through 1987 should provide an adequate amount of acceptable data for model testing.

In addition to a continuation of the present sampling program, pesticide leaching studies to resolve the following are anticipated in the 1986 and 1987 crop years:

- Soil - identification of aldicarb lateral plume movement from the band of application.
- Soil - evaluation of various size soil sampling probes for the recovery of aldicarb residues.
- Water - additional saturated zone sampling to characterize any aldicarb movement in the ground water.

Unfortunately, the 1984 analytical results were not available before the 1985 cropping year started and, therefore, any needed modifications could not be made. They are now available, and as soon as the data are processed, simulations using these data will be made. It is hoped that the 1985 data will also be available before the 1986 cropping year begins.

STEP 6d: SAMPLE ANALYSIS AND ANALYTICAL QUALITY ASSURANCE

The Beltsville Chemistry Laboratory of the Office of Pesticide Programs of the EPA was responsible for screening water and soil samples for aldicarb during the 1984 cropping year. The Analytical Chemistry Branch of the Athens Environmental Research Laboratory was responsible for the analysis of metolachlor on filter discs. A small number of peanut plant samples were also considered to evaluate plant uptake. To assure accurate and reliable analytical data, a specific quality assurance protocol was established [1] to be monitored by the Environmental Systems Branch of the Athens laboratory.

Samples arriving at the Beltsville facility were shipped in ice. As samples were received, they were logged in a book dedicated to the project and secured under lock and key. Samples remained frozen until time of analysis. All reserve sample extracts were stored frozen until disposition was determined. Sample number identification was assigned at Athens ERL. A common notebook was provided in which all entries were made. All raw data (chromatograms, mass spectra, etc.) were also collected and kept in a file dedicated to the project.

STEP 6e: COMPARE MODEL PERFORMANCE WITH ACCEPTANCE CRITERIA

At the time of this writing, the data from 1984 are still being processed, and 1985 sampling is still in progress. Therefore, although the model has been calibrated from the preliminary data, no testing or validation attempts have been made at this point. Comparisons will be made between the hydrologic and chemical outputs separately. The predicted and observed soil-water content will be compared as will be evaporation losses. If output is outside acceptance limits, an attempt at fine tuning the calibration will be made. If this does not bring the predicted values within acceptance limits, the structure of the model itself will need to be investigated. Similarly, the amount of pesticide in the soil compartments and in the soil water will be compared with model output. Also, if model predictions do not agree with the acceptance criteria (factor of two), the model will be fine tuned, and, this failing, modification in the model will need to be made.

REFERENCES

1. Cooper, S. C., R. F. Carsel, C. N. Smith, and R. S. Parrish. "Pesticide Migration in the Unsaturated and Saturated Soil Zones. Part I. A Field Study to Support Model Development and Testing, Dougherty Plains, Georgia" (Athens, GA: U.S. EPA, 1985).

2. Carsel, R. F., C. N. Smith, L. A. Mulkey, J. D. Dean, and P. P. Jowise. 1984. "User's Manual for the Pesticide Root Zone Model (PRZM): Release 1," EPA-600/3-84-109 (Athens, GA: U.S. EPA, 1984).

3. Wilson, L. G. "Monitoring in the Vadose Zone: A Review of Technical Elements and Methods," EPA-600/7-80-140 (Las Vegas, NV: U.S. EPA, 1980).

4. Owen, V. Jr. "Geology and Ground-Water Resources of Lee and Sumter Counties, Southwest Georgia," Water Supply Paper 1666, U.S. Geological Survey (1963).

5. Farm Chemical Handbook (Willoughby, OH: Meister Publishing Co., 1981).

6. Martin, H., Ed. Pesticide Manual, 2nd edition (British Crop Protection Council, 1971).

7. Richey, F. A., Jr., W. J. Bartley, and K. P. Sheets. "Laboratory Studies on the Degradation of [the Pesticide] Aldicarb in Soils," J. Agric. Food Chem. 25:47-51 (1977).

8. Coppedge, J. R., D. A. Lindquist, D. L. Ball, and H. W. Dorough. "Fate of 2-methyl-2-(methylhics) propion-aldeleyde O-methyl-carbonyl oxide (Temik) in Cotton Plants and Soil," J. Agric. Food Chem. 15:902-910 (1967).

9. Bull, D. L., R. A. Stokes, J. R. Coppedge, and R. L. Rodgers. "Further Studies of the Fate of Aldricarb in Soil," J. Econ. Entomol. 63:1283-1289 (1970).

10. Campbell, W. V., D. A. Mount, and B. S. Heming. "Influence of Organic Matter Content of Soils on Insecticidal Control of the Wireworm," J. Econ. Entomol. 64:41-44 (1971).

11. Supak, J. R. "The Volatilization, Degradation, Adsorption and Desorption Characteristics of Aldicarb [2-methyl-2-(methylthio propionaldehyde O-methylcarbomyl oxime] in Soils and Clays," Ph.D. Thesis, Texas A & M University, (1972).

12. Kearby, W. H., C. D. Ercogovich, and M. Bliss, Jr. "Residue Studies on Aldicarb in Soil and Scotch Pine," J. Econ. Entomol. 63:1317-1318 (1970).

13. Andrawes, N. R., W. P. Bagley, and R. A. Henett. "Fate and Carryover Properties of Temik Aldicarb Pesticide," J. Agric. Food Chem. 19:727-730 (1971).

14. Quarishi, M. S. "Edaphic and Water Relationship of Aldicarb and its Metabolites," Canad. Entomol. 104:1191-1196 (1972).

15. Hornsby, A. G., P. S. C. Rao, W. B. Wheeler, P. Nkedi-Kizza, and R. L. Jones. "Fate of Aldicarb in Florida Citrus Soils: 1. Field and Laboratory Studies," Proceedings of the NWWA/U.S. EPA Conference on Characterization and Monitoring of the Vadose (Unsaturated) Zone, Las Vegas, NV (1983).

16. Jones, R. L., P. S. C. Rao, and A. G. Hornsby. "Fate of Aldicarb in Florida Citrus Soils: 2. Model Evaluation," Proceedings of the NWWA/U.S. EPA Conference on Characterization and Monitoring of the Vadose (Unsaturated) Zone, Las Vegas, NV (1983).

17. Jones, R. L., and R. C. Back. "Monitoring Aldicarb Residues in Florida Soil and Water," Environ. Toxicol. Chem. 3:9-20 (1984).

18. "Clinical Scientific and Microeconomic Review of Aldicarb," U.S. Environmental Protection Agency, Washington, DC (1975).

19. Ciba-Geigy Corporation. "Dual Herbicide," Technical Bulletin, Agricultural Division, Greensboro, NC (1980).

20. Bresler, E., and R. E. Green. "Soil Parameters and Sampling Scheme for Characterizing Soil Hydraulic Properties of a Watershed," Technical Report No. 148, University of Hawaii (1982).

21. Carsel, R. F., L. A. Mulkey, M. N. Lorber, and L. B. Baskin. "The Pesticide Root Zone Model (PRZM): A Procedure for Evaluating Pesticide Leaching Threats to Ground Water," Ecolog. Model. 30:49-69 (1985).

22. Rao, P. S. C., K. S. V. Edwardson, L. T. Ou, R. E. Jessup, P. Nkedi-Kizza, and A. G. Hornsby. "Spatial Variability of Pesticide Sorption and Degradation Parameters," in ACS Symposium Series, R. Honeycutt, Ed. (Washington, DC: American Chemical Society, 1986).

23. Smith, C. N., R. A. Leonard, G. W. Langdale, G. W. Bailey. "Transport of Agricultural Chemicals from Small Upland Piedmont Watersheds," EPA-600/3-78-056 (Athens, GA: U.S. EPA 1978).

24. Smith, C. N., D. S. Brown, J. D. Dean, R. S. Parrish, R. F. Carsel, and A. S. Donigian, Jr. "Field Agricultural Runoff Monitoring (FARM) Manual," EPA-600/3-85-043 (Athens, GA: U.S. EPA, 1985).

25. Smith, C. N., and R. F. Carsel. "A Stainless Steel Suction Lysimeter for Monitoring Pesticides in the Unsaturated Zone," Soil Sci. Soc. Amer. J. 50:263-265 (1986).

CHAPTER 5

EXAMPLE MODEL TESTING STUDIES

A. S. Donigian, Jr.
Aqua Terra Consultants, Palo Alto, California

P. S. C. Rao
University of Florida, Gainesville, Florida

INTRODUCTION

Most model testing efforts with pesticides have not had the advantage of a comprehensive data collection program such as the Dougherty Plain Field Study which was specifically designed to provide a reasonably adequate data base for model testing and validation. Although numerous field studies have provided and are currently providing data on pesticide fate in the vadose zone, much of the available data are of limited use for model evaluation. A frequent mismatch occurs between the needs of the experimentalist who designs the field studies and the needs of the modeler for model testing [1]. This manual is aimed at helping bridge this gap by informing and educating these two groups regarding the data needs and problems of each group so that future data collection efforts will more fully meet the needs of both the experimentalist and the modeler.

At present most model testing and validation studies must rely on a limited data base of a few field studies designed to collect data for a variety of purposes. This section discusses and summarizes a few selected model testing studies of this type which relied on available field data. These studies demonstrate the types of comparisons that are often made between field data and model predictions, the procedures required for model testing, and the model performance or acceptance criteria (or statistical tests) used to quantify model performance. Since previously collected data are utilized, this section demonstrates only steps 4 and 6e in the field validation procedures discussed in Chapter 3.

The remainder of this chapter is based primarily on two recent studies aimed at testing and evaluating the three models discussed in Chapter 1. The first, by Watson and Brown [2], involved a comprehensive evaluation of the SESOIL model. They compared SESOIL model predictions with published field data for aldicarb nematicide and with the predictions by Jones et al. [3] of three other models (PRZM, PESTAN, and PISTON). The second study reviewed here is that of Carsel et al. [4] who evaluated the PRZM model using field data for metalaxyl pesticide. We will summarize the findings of these two studies here in order to illustrate the problems encountered in testing models using field data, and those encountered in the testing methodologies used to evaluate model performance. There are also a number of other on-going model testing studies aimed at comparing PRZM with other, more research-oriented, models.

TESTING METHODOLOGY USED BY WATSON AND BROWN

A two-tiered approach was used by Watson and Brown [2] to test SESOIL. The first tier involved making comparisons between SESOIL predictions and results from an analytical solution to the partial differential equation describing solute transport in the unsaturated zone; a closed-form solution to the one-dimensional, convection-dispersion equation was used [5]. The second tier of testing involved comparing SESOIL predictions with field data and with predictions of other models tested with the same field data sets. Comparisons were made with two types of data sets. The first type was data collected on chemical leaching in the unsaturated zone. The second type was data collected during pesticide runoff studies on field-scale watersheds. Only the procedures and results of the chemical leaching comparisons will be presented in this section; the interested reader is referred to the original report for the results of the first-tier SESOIL testing (i.e., comparisons with the analytical solution) and the pesticide runoff comparisons. Chapter 1 summarized the primary conclusions of the entire SESOIL model testing effort.

Jones et al. [3], identified four major types of predictions that should be provided by models designed for describing pesticide fate in the unsaturated zone and for describing the potential for contamination of ground water. These are listed as follows in decreasing order of importance:

1. the transit time for the pesticide within the unsaturated zone
2. pesticide mass emissions into the ground water
3. pesticide concentration in the ground water
4. pesticide concentration distribution within the soil profile

The testing by Watson and Brown emphasized the first two types of comparisons, transit time, and mass emissions (or loading) to ground water. They also included comparisons of the pesticide mass remaining in the soil profile at different times following application which are a useful aggregate measure of transit time, loading to ground water, and degradation processes. Comparisons of pesticide concentrations in ground water was beyond the scope of the models and testing effort. Comparisons of pesticide concentrations in the soil profile are also difficult tests due to spatial and temporal variations in soil characteristics and processes as discussed in Chapters 1 and 11. Moreover, the available field data were not sufficiently detailed to allow an assessment of spatial variations.

The tests by Watson and Brown were initially conducted with limited calibration of SESOIL parameters since Bonazountas and Wagner [6,7] suggest that the model can provide reasonable results with limited calibration. Tests involving calibration and verification were also conducted with those data sets having sufficiently long records. A split-sample calibration/verification procedure [8] was used when sufficient data were available. A split-sample procedure involves dividing the data set into two separate data sets: one of which is used to calibrate model parameters, and the other is used to test or verify model parameters.

EPA-OTS currently uses SESOIL primarily to predict the extent to which a chemical will leach to ground water. For this reason, the first tier of testing by Watson and Brown focused entirely on examining the unsaturated zone transport algorithms in SESOIL. An analytical solution to the one-dimensional, convection-dispersion equation was selected for testing purposes because: (1) model behavior for a range of hydrologic conditions and chemical characteristics could be examined rapidly, and (2) model behavior under relatively idealized conditions could be examined. The latter reason is perhaps the most important. Analytical solutions are generally derived assuming idealized (i.e., homogeneous and isotropic) soil properties, a constant pore-water velocity (i.e., steady water flow), and idealized chemical characteristics (e.g., equilibrium, linear, reversible adsorption, and first-order degradation). Comparisons of model predictions with analytical solutions are often used to examine and evaluate the numerical procedures and sensitivity of model parameters [9]. By parameterizing SESOIL and the analytical model to represent the same conditions, similarities and dissimilarities in model behavior were identified.

FIELD DATA FOR MODEL TESTING

Field data sets useful in testing a model, such as SESOIL, are limited in number for the following reasons:

1. Most field studies focus on either chemical leaching in the unsaturated zone or runoff losses from watersheds; few field studies have measured both.

2. Because of the high costs associated with sampling and analysis, most field studies of chemical leaching are of short duration.

3. Since most field studies are conducted for purposes other than for generating data to test models, they are often incomplete.

Watson and Brown initially screened available data sets published in the literature and identified seven candidate data sets for model performance testing: (1) aldicarb leaching in Florida citrus groves [10]; (2) aldicarb pollution of ground water on Long Island, New York [11]; (3) a field study of the transport of agricultural chemicals from the small upland piedmont watersheds near Watkinsville, Georgia [12]; (4) a field study of agricultural chemical transport in the Four Mile Creek watershed in Iowa [13]; (5) aldicarb leaching in Wisconsin [14]; (6) a field study of bromide movement in California [15]; and (7) radionuclide movement in soils near Hanford, Washington [16].

The above data sets were selected for consideration for several reasons. First, all of the data sets are fairly complete in terms of both hydrologic and chemical data. Second, measurements of chemical movement in the unsaturated zone were made in all cases, and chemical losses due to runoff and soil erosion were measured in two of the studies (i.e., Watkinsville and Four Mile Creek). Third, the initial mass of chemical (i.e., the source term) is reasonably well defined; this is particularly true for those field studies involving pesticides. Finally, other model tests have been conducted with all of the data sets. The Florida and Wisconsin aldicarb data sets were chosen by Watson and Brown for chemical leaching tests; only the Florida data set is briefly discussed below. The original report by Watson and Brown [2] includes summary descriptions of the other data sets listed above.

The Florida field study was undertaken to assess the processes which control the movement of aldicarb in citrus groves and to compare field and laboratory-derived degradation and sorption parameters. Hornsby et al. [10], describe laboratory and field experiments conducted in two citrus groves, and Jones et al. [3], describe the model performance tests conducted with the data. Field and modeling efforts concentrated on: (1) the transit time for the aldicarb total toxic residues (TTR) within the unsaturated zone, (2) aldicarb TTR mass emissions from the vadose zone, (3) aldicarb concentrations in the ground water, and (4) aldicarb distributions within the soil profile.

One field site was in Seminole County near Oviedo, Florida. The site is a bedded grove composed primarily of Immokalee fine sand and Delray fine sand. Hornsby et al. [10], note that the Immokalee series soils are poorly drained, coarse textured, and strongly acid. The Delray series soils are deep, poorly drained or very poorly drained sands with a thick, highly organic surface layer. The second field site is located in Polk County near Lake Hamilton, Florida. The soil at this site is a well-drained, deep coarse sand, typical of the central ridge area of Florida.

Aldicarb was applied to the Oviedo and Lake Hamilton sites on February 15 and 16, 1983, respectively. Soil samples were collected on March 3 and 4, April 5 and 6, May 2 and 3, June 14 and 15, and August 23 and 24. Soil samples at the field sites were collected every month throughout 1983; however, data for a 4-month period following application were available for model testing. Samples were collected in 30-cm intervals to a depth of 150 cm at the Oviedo site, in 30-cm intervals to a depth of 60 cm, and in 60-cm intervals to a depth of 300 cm at the Lake Hamilton site. Laboratory studies were conducted on undisturbed soil core samples to obtain soil-water characteristic curves, soil bulk densities, and saturated hydraulic conductivities. Laboratory studies were also conducted using bulk soil samples from different horizons to obtain the data required to estimate sorption coefficients and degradation rate coefficients.

Aldicarb concentrations in soil samples were found to be highly variable. Despite this, the data did show detectable concentrations of aldicarb progressing towards the water table with time [10]. Sampling was continuous throughout the soil profile, but reported TTR concentrations are the result of composite sampling over 30 or 60 cm intervals. Jones et al. [3], tested three different models using these data: (1) PISTON [17], (2) PESTAN [18], and (3) PRZM [19].

As noted above, the Florida and Wisconsin aldicarb data sets were selected by Watson and Brown [2] for the second tier of model testing because they are the most complete of all the chemical leaching data sets reviewed. They provided the primary test of unsaturated zone hydrology and pollutant transport cycles in SESOIL. However, the short time frame covered by these data sets (i.e., four to six months) precluded the use of a split sample calibration and verification procedure. Therefore, SESOIL was tested by Watson and Brown with only limited calibration. Any additional parameter adjustments required to bring SESOIL results in line with the field data were evaluated for reasonableness.

QUANTITATIVE MEASURES OF PERFORMANCE

In evaluating the performance of a model relative to some field measurement or another model, qualitative rather than quantitative statements are often made. Statements like "good or reasonable agreement was achieved between the model and the observed results" are commonplace. Such statements do not provide a sound basis for judging model accuracy. Dissatisfaction with these types of qualitative statements have prompted researchers to begin to develop or adapt quantitative, statistical measures that can be used to generate more meaningful statements.

Three general procedures have been identified that can be used to provide quantitative measures of performance for a model [8]. These procedures include:

1. Paired-data performance: the comparison of simulated and observed data for exact locations in time and space.

2. Time and space integrated, paired-data performance: the comparison of spatially and temporally averaged simulated and observed data.

3. Frequency domain performance: the comparison of simulated and observed frequency distributions.

In selecting one or more of these procedures, it is important to consider the characteristics of the "simulated" and "observed" data. For example, in this case SESOIL generates the simulated data by calculating seasonally averaged fluxes of water and pollutant mass loading rates. It also calculates compartmentally-averaged pollutant concentrations at the end of each season.

The observed field data tend to be for specific points in time and generally represent either points in space or regions (e.g., the upper 15 cm of soil). Since it is impossible to spatially or temporally disaggregate the simulated data (i.e., SESOIL results), the observed data have to be aggregated. Another characteristic of both the simulated and observed data is that they tend to cover relatively short periods of time (say 4 to 24 months). As a result, it is difficult to develop a meaningful characterization for concentration variations with time (e.g., frequency distributions). Given these characteristics, a time and space integrated, paired-data performance procedure by Watson and Brown [2] was selected and applied to all comparisons.

A number of statistical techniques can be used to obtain a quantitative measure of the difference between simulated and observed data. Moore et al. [20], discuss different statistical measures and associated aggregate statements useful for evaluating the performance of air quality models. Ambrose and Roesch [21] discuss several measures used to evaluate an estuary water quality model.

Basically, two classes of statistical models were selected to generate quantitative measures of performance for the SESOIL tests. The first class is standard linear regression statistics. As Ambrose and Roesch [21] note, regression statistics for observed and simulated data can be calculated by:

$$O^i = a P^i + b \tag{5-1}$$

where O^i = observed data

P^i = simulated data

a = slope

b = intercept

The slope, a, provides a useful measure of the accuracy of model results; values less than 1.0 imply over-prediction by the model, while values less than 1.0 imply under-prediction. The intercept, b, is an indicator of statistical errors. Perfect agreement between the observed data and model predictions occurs when a = 1 and b = 0. The precision of model results can be measured by r, the correlation coefficient. It should be recognized that an implicit assumption made in this analysis is that the predicted values (P^i) are "fixed" and are without error. This assumption arises from the deterministic nature of the models used, where a single value is assigned for each of the model parameters resulting in a unique predicted value (see Chapter 1). Since most, if not all, model parameters are spatially-variable, predicted values also have certain error. Thus, computation and interpretation of the linear regression parameters, as shown in eq. (5-1), may not be strictly valid. Under such conditions, to estimate parameters in linear models Halfon [22] has recently recommended the use of geometric mean functional regression (GMF), which takes into account the errors in both the dependent and the independent variables (in this case, O^i and P^i).

The second class of methods are typically called estimation of error statistics. Ambrose and Roesch [21] state that the average error, $\overline{E}$, and its associated relative error, RE, can be used to measure accuracy and systematic errors:

$$\overline{E} = \frac{1}{N} \sum_{i=1}^{N} (P^i - \overline{O^i}) \tag{5-2}$$

$$RE = \overline{E}/\overline{O} \tag{5-3}$$

where $\overline{O}$ = observed data mean

N = number of data points

The average error gives the absolute amount by which a given quantity is over- or under-predicted, while the relative error gives the percentage of over- or under-prediction.

Other estimation of error statistics that can be calculated include the standard error of estimate, SE, and its coefficient of variation CV, as follows:

$$SE = \left[\frac{1}{N} \sum_{i=1}^{N} (P^i - O^i)^2\right]^{1/2} \tag{5-4}$$

$$CV = \frac{SE}{\overline{O}} \tag{5-5}$$

The standard error of estimate is the difference between the actual observed and predicted values, while the coefficient of variation of standard error gives the average relative difference.

Parameter Estimation and Model Calibration

A two step approach was used to estimate and calibrate SESOIL model parameters. Data from the literature, guidance provided in the SESOIL manual, and chemical property parameter estimation methods provided the basis for an initial set of input parameters. SESOIL results based on this parameter set were then compared with observations on depth-averaged soil-water contents, percent aldicarb TTR leached to ground water, and percent aldicarb TTR remaining in the soil profile over time.

The second step involved model calibration by adjusting selected input parameters until reasonable agreement with field data was achieved. Soil-water content was again the key calibration end point used for the hydrologic cycle. Evaporation was not measured at either site, and runoff was minimal. Percent aldicarb TTR leached to ground water was the end point used to calibrate the pollutant cycle parameters; the high degree of variability in measured vertical distributions of aldicarb TTR concentrations in the soil profile precluded the use of these data for model calibration. As will be discussed later, however, these data were useful in a qualitative sense for interpreting the performance of SESOIL.

The initial hydrologic cycle parameter values for the Lake Hamilton site are listed in Table 5-1. As was discovered by Watson and Brown [2], the intrinsic permeability and saturated hydraulic conductivity suggested in Table ID-1 of the SESOIL user's manual were not consistent. Using the suggested saturated hydraulic conductivity values, SESOIL predicted soil-water contents that were too low compared to field measurements; using a conductivity value derived from the suggested intrinsic permeability value produced soil-water contents that were too high.

Table 5-1. Initial and Calibrated Hydrologic Cycle Parameter Values for the Lake Hamilton, Florida Site.

Parameter	Initial Value	Calibrated Value
Hydraulic Conductivity, cm/sec	0.1, 0.001	0.01
Disconnectedness Index, c	3.7	4.1
Porosity, percent	35	43

Agreement was improved by increasing the porosity and the disconnectedness index (see Table 5-1) and by adjusting the saturated hydraulic conductivity to a value of 0.01 cm/sec; this is equivalent to the laboratory results obtained by Hornsby et al. [10]. Figure 5-1 shows the resultant SESOIL prediction compared to measured depth-averaged soil-water contents. The error bars show the minimum and maximum average measured soil-water contents. The calculated estimation of error statistics show that on the average SESOIL over-predicted soil-water contents by 0.43 percent (E) with a relative error (RE) of 0.07 (Table 5-2). The standard error (SE) between the predicted and observed soil-water contents was 0.88 percent.

The same initial set of parameter values was used for the Oviedo site. Attempts to calibrate SESOIL and to keep parameters in a reasonable range based on literature data proved unsuccessful. In examining the measured data further and in discussions with Dr. A. G. Hornsby of the University of Florida, it was discovered that there was one or more low permeability layers in the soil column. These layers restrict drainage and produce highly variable soil-water distributions. Since SESOIL is unable to represent the effect of such layering on water movement, no further attempts were made to calibrate the model on the Oviedo site.

An initial set of pollutant cycle parameters was estimated using published literature data on aldicarb migration and fate in soils and estimation techniques in Lyman et al. [23]. The values are shown in Table 5-3.

Using this initial parameter set, SESOIL predicted that 65.6 percent of the total mass of aldicarb TTR applied to the site would leach to ground water during the growing season while Jones et al. [3], reported that about 4-8 percent of the applied aldicarb TTR had leached to ground water at the Lake Hamilton site. Runoff and volatilization losses were predicted to be minimal. The predicted runoff losses are consistent with those observed by Hornsby et al. [10]; low volatilization losses are also consistent given the chemical properties of aldicarb.

Calibration of the pollutant cycle parameters focused on adjusting K_{oc} and hydrolysis rate constants until the predicted mass leached to ground water decreased within the measured range of 4-8 percent. With a K_{oc} of 240 and a hydrolysis rate of 0.015/day, SESOIL predicted that 10 percent would reach ground water. By increasing the hydrolysis rate to 0.0254/day, 4.3 percent was predicted. The calibrated K_{oc} parameter value is questionable given values reported in the literature. Hornsby et al. [10], found that K_{oc} values for the Lake Hamilton site were between 0 and 47 cm^3/gm. With the exception of the value found by Hough et al. [24], for a clay soil, ranges published by Bromilow et al. [25], Supak [26], and Hornsby et al. [10], are roughly an order of magnitude lower than the calibrated value. The calibrated hydrolysis rate is comparable with data presented by Smelt et al. [27,28], and Bromilow et al. [25], for similar soil, pH, and temperature conditions. It

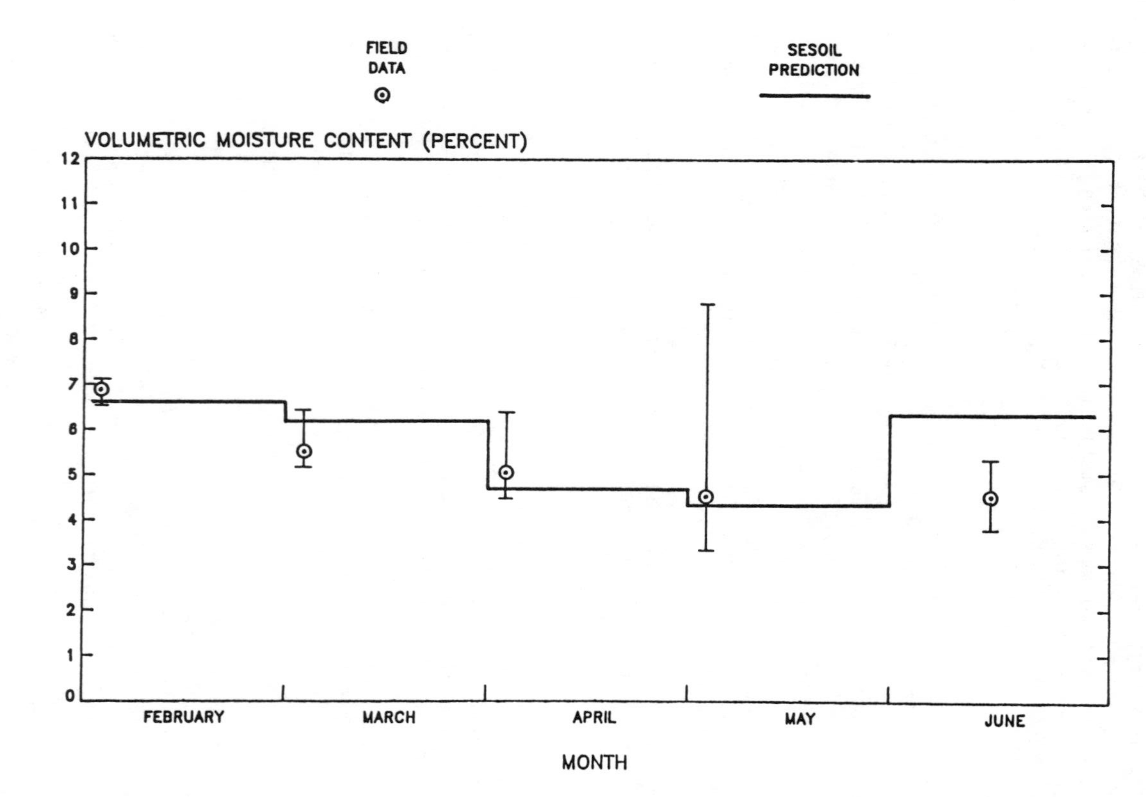

Figure 5-1. Comparison between SESOIL predicted and measured depth-averaged soil-water contents for the Lake Hamilton site.

Table 5-2. Computed Statistics for Comparisons at the Lake Hamilton Site.

Comparison	$O^i = mP^i + b$						
	m	b	r^2	$\overline{E}$	RE	SE	CV
SESOIL/Observed Volumetric Soil-Water Content, Percent	0.58	2.02	0.38	0.43	0.07	0.88	0.17
SESOIL/Observed Aldicarb TTR Residues, Percent	0.88	-10.22	0.77	16.60	0.47	23.46	0.66

is at the high end of the range reported by Hornsby et al. [10], for the Lake Hamilton site. Because of the extreme parameter values obtained in this "calibration" exercise, it may be concluded that SESOIL does not adequately represent the aldicarb behavior at this field site.

Figure 5-2 shows a comparison between predicted and observed total aldicarb residues in the soil profile using the calibrated parameter values in Table 5-3. On the average, SESOIL predicts that 16.6 percent more of aldicarb TTR remains in the soil column (E); the average difference (SE) in predicted and observed values

Table 5-3. Initial and Calibrated Pollutant Cycle Parameter Values for the Lake Hamilton, Florida, Site.

Parameter	Initial Value	Calibrated Value
Solubility, mg/l	6000	6000
K_{oc}, cm^3/gm	39	240
Diffusion coefficient in air, cm^2/sec	0.06	0.06
Biodegradation rate, /day	0.0	0.0
Henry's Law constant, $M^3atm/mole$	0.33E-08	0.33E-08
Neutral hydrolysis rate, /day	0.0051	0.025

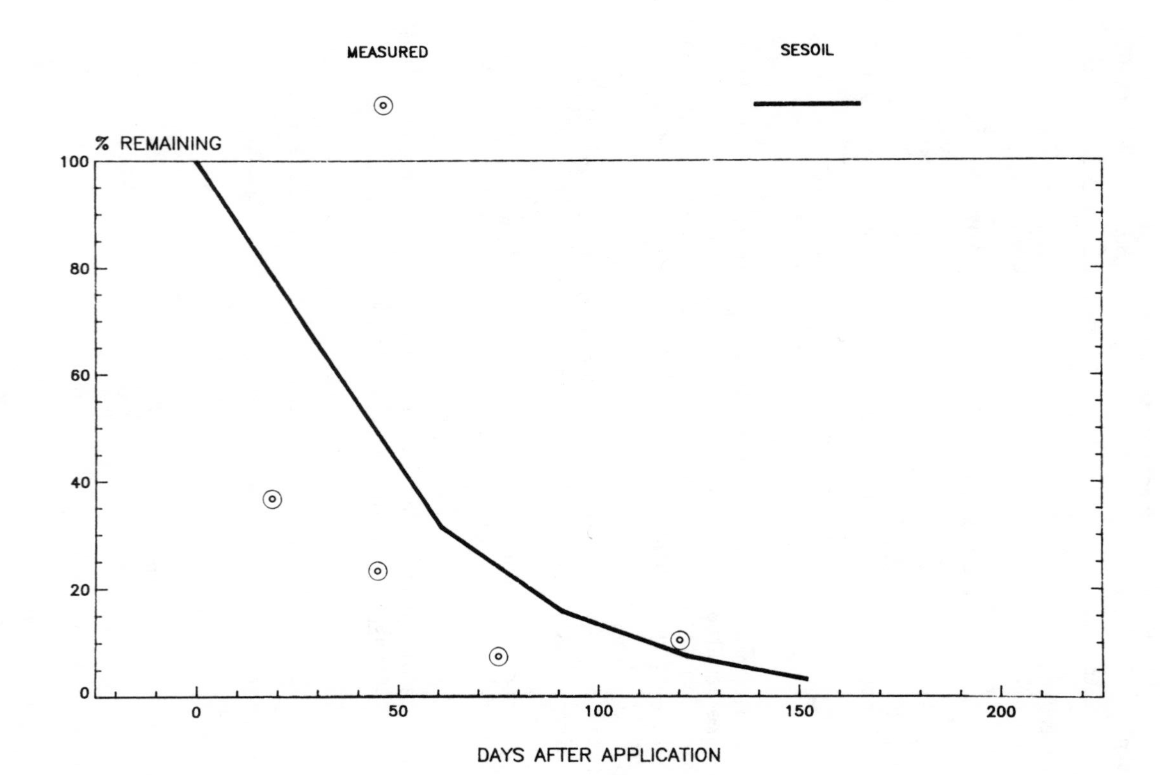

Figure 5-2. Comparison of SESOIL predicted and measured percent aldicarb remaining in the soil column using calibrated parameter values for the Lake Hamilton, Florida, site.

is 23.5 percent. See Table 5-2 for values of other calculated performance measures.

Figures 5-3, 5-4, and 5-5 show predicted and observed vertical variations in aldicarb TTR concentrations with time. Again, the high K_{OC} value required to limit the rate of leaching to ground water precludes the aldicarb TTR from migrating vertically to any large extent. The field data, however, again show a distinct downward progression of the aldicarb TTR pulse.

Comparison with Other Models

Jones et al. [3], report on a model evaluation study that was conducted using the data collected by Hornsby et al. [10]. Three models, PESTAN, PISTON, and PRZM, were run using equivalent model parameters, and the simulation results were compared with the observed mass of aldicarb leached to ground water and total residual aldicarb remaining in the soil column over time. A comparison between SESOIL results and those of Jones et al. [3], was obtained by using an equivalent set of pollutant cycle parameters in SESOIL (see Table 5-4). The hydrologic parameters were the same as those obtained in calibrating SESOIL against the measured soil-water content data.

Table 5-5 lists the percent of total applied aldicarb TTR leached to ground water predicted by each of the models. The PISTON and PRZM results are within the 4-8 percent range measured by Hornsby et al. [10]. The PESTAN model under-predicted the mass leached to ground water while SESOIL overpredicted by a factor of 4-10.

A comparison of predicted and measured aldicarb TTR residues in the soil profile with time is given in Figure 5-6. This figure shows that PESTAN, PISTON, and PRZM give comparable results while SESOIL initially predicts higher residues and later predicts lower residues. A comparison of quantitative measures of performance between PRZM and the observed residues, and between SESOIL and the observed residues shows that PRZM had only slightly better agreement than SESOIL (see Table 5-6). The estimation of error statistics are similar for both models.

The increase in the SESOIL-predicted rate of aldicarb TTR dissipation around Day 30 corresponds to the time when aldicarb is initially released to ground water. If this release had not occurred, SESOIL would have overpredicted soil residues given the initial rate of dissipation predicted prior to Day 30. In checking this rate of dissipation, it was discovered that it was about <u>two times smaller</u> than the hydrolysis rate input to the model. An error in the hydrolysis algorithm, discussed by Watson and Brown [2], was later found to be the source of the problem. Thus, while agreement between SESOIL predicted and observed soil residues is qualitatively good, it is only because of the over-prediction of aldicarb leaching to ground water.

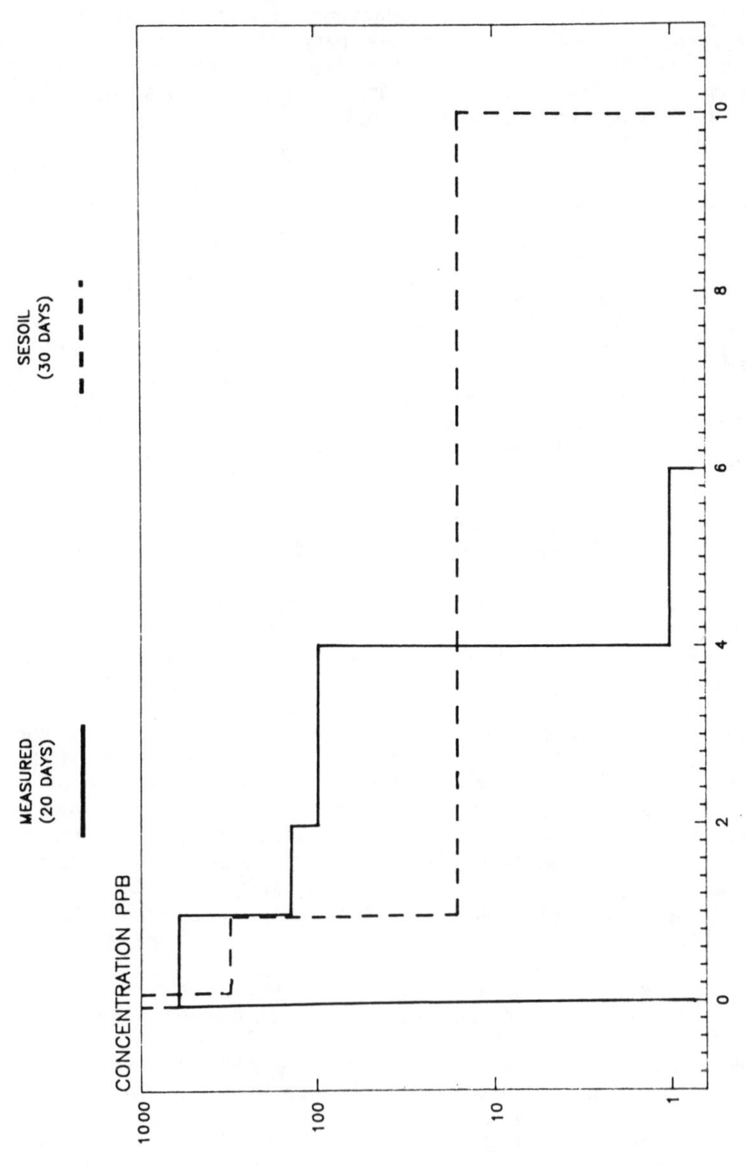

Figure 5-3. Comparison of SESOIL predicted and measured vertical concentration distributions 30 and 20 days after application, respectively, using calibrated parameter values for the Lake Hamilton, Florida, site.

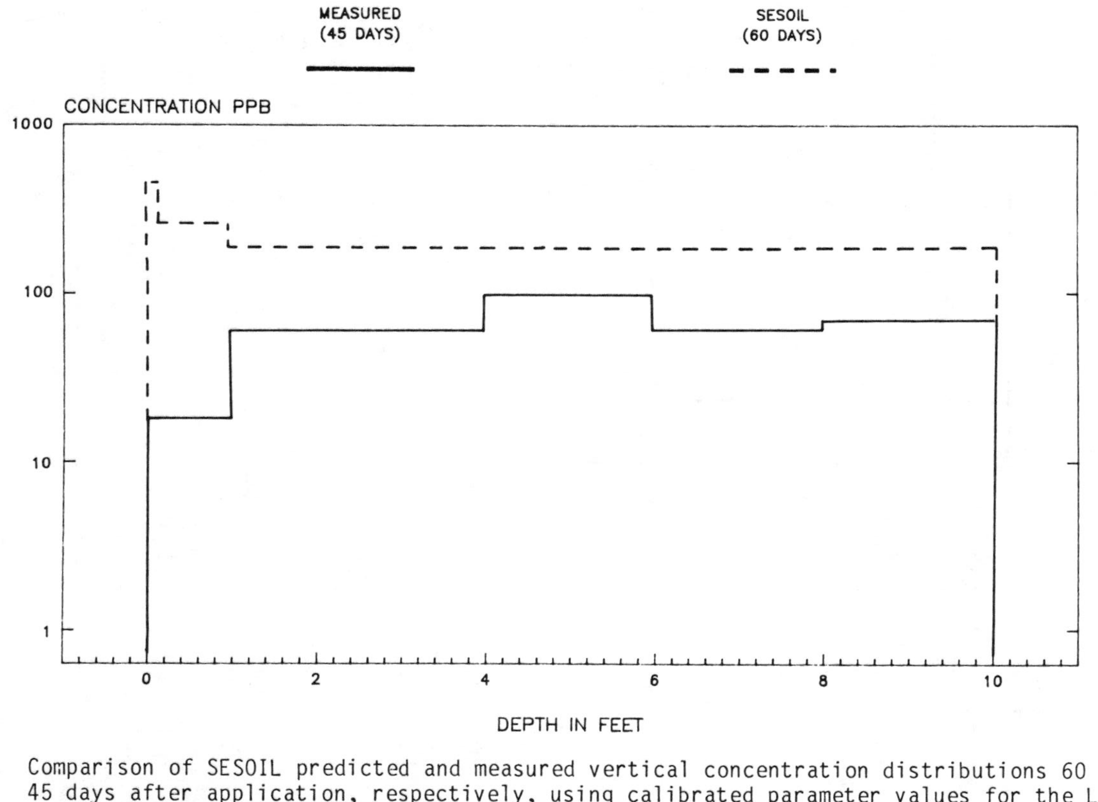

Figure 5-4. Comparison of SESOIL predicted and measured vertical concentration distributions 60 and 45 days after application, respectively, using calibrated parameter values for the Lake Hamilton, Florida, site.

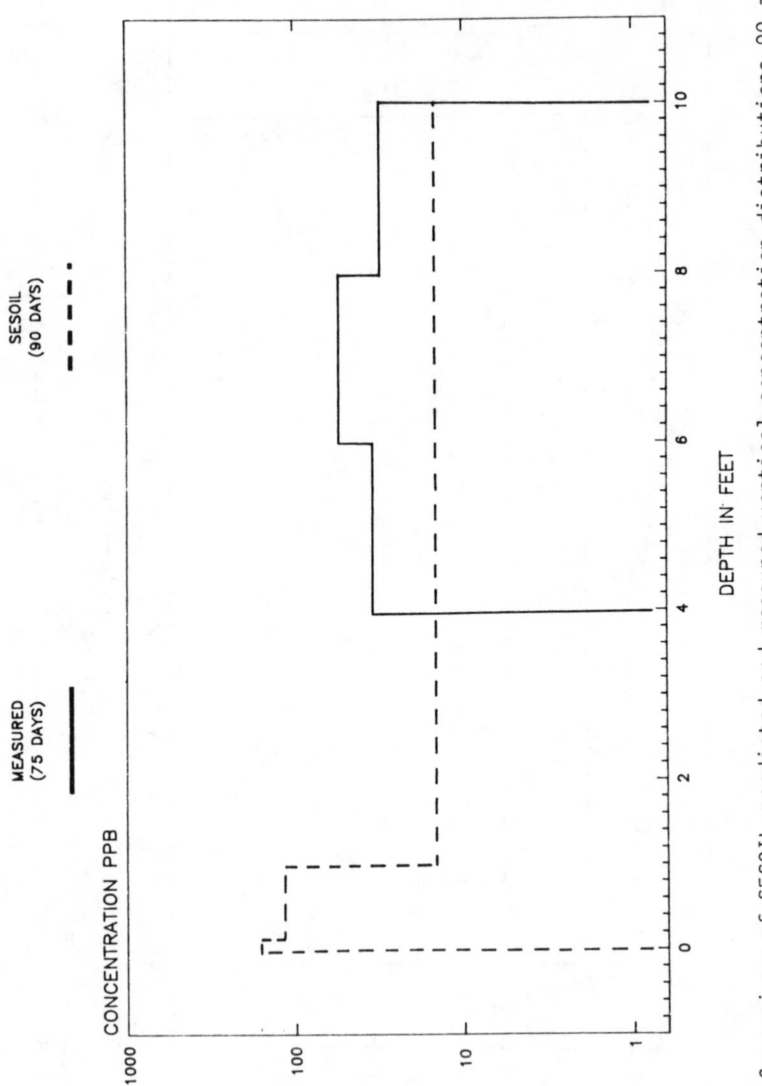

Figure 5-5. Comparison of SESOIL predicted and measured vertical concentration distributions 90 and 75 days after application, respectively, using calibrated parameter values for the Lake Hamilton, Florida, site.

Table 5-4. Pollutant Cycle Parameters Used in Testing PRZM, PISTON, and PESTAN on the Lake Hamilton, Florida, Site [3].

Parameter	Value
K_{oc}, cm^3/gm	26.7
Neutral hydrolysis rate, /day	0.019

The new version of the SESOIL code was also tested using the Lake Hamilton calibrated parameter values. The new version predicted that slightly less aldicarb TTR would leach to ground water: 4.0 percent as opposed to 4.3 percent. The addition of a fourth compartment and modifying the model to retain chemical in each compartment until the pollutant penetration time was exceeded produced only a small reduction in the amount of chemical leached to ground water.

The predicted chemical distribution in the soil column was also only slightly improved. Aldicarb was still retained in the upper layers because of the high K_{oc} value needed to limit leaching to ground water. Aldicarb concentrations in the bottom compartment were slightly lower during the first couple of months and slightly higher during the last two months.

PRZM Testing by Carsel et al.

Carsel et al. [4], evaluated the PRZM model performance using field data for metalaxyl pesticide collected during the 1980-81 period at two field sites, one each in the tobacco growing areas of Florida and Maryland. The soils at the 3.5-ha field in Florida are

Table 5-5. Comparison of Model Leaching Predictions for Lake Hamilton.

	PESTAN	PISTON	PRZM	SESOIL
Percent of Applied Leaching below 10 feet	0-0.2*	7.4	3.5	35.4

*Results depend on the value chosen for the curve coefficient. The lower number represents the model value provided for sand; the higher number is calculated based on the curve coefficient which gives a soil water content approximately equal to the field capacity.

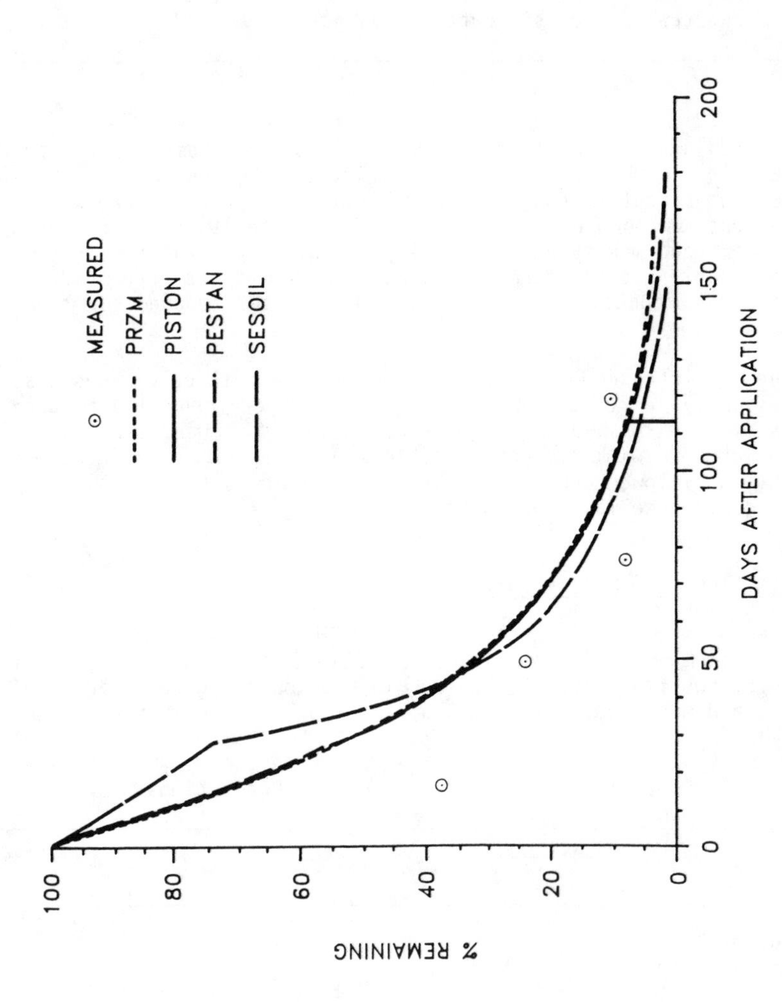

Figure 5-6. Comparison of PRZM, PISTON, PESTAN, and SESOIL predicted and measured percent aldicarb remaining in the soil column using parameter values from Jones et al. [3].

Table 5-6. Revised Statistics for Model Comparisons at the Lake Hamilton Site.

Comparison	$O^i = mP^i + b$						
	m	b	r^2	E	RE	SE	CV
SESOIL/Observed Aldicarb Residues, Percent Remaining	0.78	-1.89	0.76	12.40	0.35	22.32	0.63
PRZM/Observed Aldicarb Residues, Percent Remaining	0.91	-6.50	0.86	10.80	0.30	17.03	0.48

classified as Blanton sand and are characterized by high permeability, low water-holding capacity, and low soil organic matter content. At the 0.6-ha Maryland site, the soils are classified as Marlton sandy loam. These soils are characterized by low permeability, large water-holding capacity, and high soil organic matter content. Hydrologic and soil characteristics of these two sites are summarized in Table 5-7.

Subsequent to metalaxyl (Ridomil 2E) application at a rate of 2.2 kg a.i./ha, tobacco was transplanted at both sites. Soil samples were taken at the Florida site in 15-cm increments to a depth of 90 cm on the day of application (0 days), and on 26, 55, 85, and 154 days following pesticide application. At the Maryland site, soil samples were collected at the same depth increments as in Florida but at 0, 15, 30, 44, 49, 61, 76, 91, 106, 121, 135, 152, 219, 261, and 287 days. For each sampling, four soil cores (one in each of the quadrants of the field site) were collected and composited and analyzed to obtain a "field-averaged" metalaxyl residue concentration with depth and time. This method differs from that used by Hornsby et al. [10], who took 4 soil cores in each of the quadrants of the field plot, analyzed all 16 replicates for aldicarb, and then computed an average pesticide concentration. It is not known which of these two methods yields a statistically valid value for the "field average" pesticide concentration.

Because soils are spatially-variable, both the numbers and locations of soil sampling have a direct influence on the estimated average value of a soil property. Compositing several soil samples only adds another element of uncertainty to the estimated value because the soil samples taken "at random" may not be statistically independent. The problems of compositing soil samples and estimating a "field-average" value of a soil property have recently been examined by Webster and Burgess [29]. They showed that the estimation variances of a soil property (e.g., pollutant concentration)

Table 5-7. Summary of Hydrologic and Soil Characteristics of the Two Field Sites Used in the Metalaxyl Study (adapted from Carsel et al. [4]).

Characteristic	Florida	Maryland
Hydrologic Characteristics*		
Precipitation, cm/yr	110	100
Runoff, cm/yr	10	25
Evapotranspiration, cm/yr	70	60
Recharge past root zone, cm/yr	30	15
Hydrologic soil group	A	C
Soil Characteristics**		
Soil Series	Blanton	Marlton
Field-capacity, (%)	9.1, 9.1, 25.7	20.7, 33.9, 25.7
Wilting-point, (%)	3.3, 3.3, 14.8	9.5, 23.9, 14.8
Organic matter, (%)	1.5, 1.0, 0.5	2.5, 1.0, 1.5
Bulk density, g/cm^3	1.45, 1.55, 1.65	1.25, 1.3, 1.3

*Average annual values.
**For each soil characteristic, three values shown are for surface soil, subsurface soil, and substratum, respectively. Depths corresponding to these zones were not specified by the authors.

in the soil samples will be essentially equal to the Kriged estimates when the following criteria are met: (1) the property is additive; (2) the semivariogram, describing spatial-dependence of the property, is either linear or spherical; (3) equal portions of the soil samples are mixed in bulking; and (4) the samples to be bulked are taken from an optimally configured sampling grid. Webster and Burgess [29] stated that the optimal configuration for sampling is to collect samples at the nodes of a centrally-placed grid with its interval $d=(L/n^{1/2})$, where L is the length of the block (or plot), and n is the number of sampling points. To illustrate these criteria, Webster and Nortcliff [30] reanalyzed the data collected by Khan and Nortcliff [31] for micronutrient (Cu, Fe, Mn, and Zn) concentrations in soil samples from a 1-ha field.

Since pesticide concentration is an additive property, requirement (1) above is satisfied. As long as soil samples taken over short distances are bulked, condition (2) may also be valid. A common practice in bulking soils is to mix nearly equal weights (or volumes) of soil; thus requirement (3) is also satisfied. Condition (4) is probably not met in most studies, and the pollutant concentration values estimated from bulk samples may be biased. Thus, in the Carsel et al. [19] field experiment, an "optimum" sample and an "unbiased" pesticide concentration estimate are obtained only if the four soil samples were collected from the <u>center</u> of each quadrant and then equal portions were mixed; otherwise the estimation variance would be larger. Apparently, the field procedures were designed to meet this requirement [32].

On the other hand, if the distance over which pesticide concentrations were spatially-dependent was shorter than the quadrant size, i.e., the samples were, in fact, spatially independent, only the number of sampling points and not their locations in the field would determine the estimation variance [29,30]. The more non-normal the population frequency distribution of the soil property of interest, the larger would be the number of samples needed to estimate the mean value with a given level of confidence (Chapter 11). From the foregoing discussion it should be evident that unless the spatial structure of the variability in the soil property is known, as determined by the semivariogram, the economic benefits gained in bulking soil samples and reducing the costs of laboratory analyses may come at the expense of increased estimation variance. Also, all information on the nature of the soil spatial variability is lost by compositing samples taken from different location within a field site. As we have seen, such knowledge may be crucial in determining the statistical validity of the sample collected and the pollutant concentration measured.

Carsel et al. [4], selected these two field sites because of the availability of general hydrologic and soil characteristic data. However, several site-specific hydrologic input data required in PRZM (e.g., daily records of rainfall, pan evaporation, runoff, ground-water recharge, etc.) were not collected and had to be estimated on the basis of model calibration using historical climatological data from nearby meteorological stations. Unavailability of site-specific data for model parameters is a major problem especially when data from field studies are used where monitoring is the primary goal rather than model testing. A similar problem is encountered in estimating pesticide-specific model parameters, such as sorption coefficient (K_D) and degradation rate coefficient (k) or half-life ($t_{1/2}$). Carsel et al. [4], estimated the K_D values based on a K_{om} value of 40. However, the k values for metalaxyl had to be estimated based on the decrease in total amount of metalaxyl remaining in the 90-cm soil profile at various times during the monitoring period. They noted that a first-order rate model with $k = 0.014$ day^{-1} described well ($R^2 = 0.99$) the metalaxyl losses at the Florida site. A biphasic first-order model was needed to describe the metalaxyl data from the Maryland site; the rate coefficient for the first phase (0-30 days) was 0.0455 day^{-1}, while that for the second phase (>30 days) was 0.00453 day^{-1}. Carsel et al. [4], cite others who have noted a similar multiphasic pesticide degradation under field conditions with this degradation depending on the timing of pesticide application and the occurrence of rainfall or irrigation events. They attribute it to various processes (e.g., volatilization, photo-degradation) that affect pesticide fate near or at the soil surface, but do not occur once the pesticide has leached to subsurface depths. It should be recognized that the rate parameters estimated in this manner are for pesticide "loss" via all pathways except leaching and represent the "depth-averaged" values for the entire soil profile sampled (i.e., 0-90 cm in this case). Therefore, another likely explanation for the biphasic degradation is the differences in degradation rates with soil depth. This points out the limitations of assigning a single degradation rate coefficient for the entire soil profile.

Observed and simulated (PRZM) metalaxyl concentration profiles for the Florida site are compared in Figure 5-7. It is evident that the measured and predicted concentration profiles agree for the 55 and 85-day sampling but do not agree for the first sampling date (26 days post-application). A linear regression of measured and simulated data yielded coefficients of determinations (R^2) of 0.33, 0.90, and 0.95 for the 26, 55, and 85 day sampling, respectively. Note that the areas under the predicted and measured curves representing the total amount of metalaxyl in the 90-cm profile agree in all cases. Since the degradation rate coefficient value was estimated by calibrating the model to measured data, this is an expected result. Thus, the failure of the PRZM model to predict the 26-day post-application metalaxyl concentration profiles may be attributed to errors in the model input data (recall that site-specific daily records for hydrologic and meteorological data were unavailable and were estimated), errors in measured data (in soil sampling and pesticide analysis), and the inadequacy of the model to simulate water and pesticide movement.

As a result of lower precipitation, a higher runoff (Table 5-7), and a larger sorption coefficient compared to the Florida site, metalaxyl did not leach past the 15-cm depth during the entire sampling period at the Maryland site. Thus, PRZM model simulations were compared with the measured data only for the 0-15 cm depth (Figure 5-8). A linear regression of measured and predicted values yielded an R^2 of 0.75, indicating a reasonable agreement. However, it should be noted that since leaching past the 15-cm depth was not a significant factor, the dominant loss pathways would be runoff, degradation, and uptake. Of these, the model was calibrated to estimate the degradation rate coefficient. Thus, agreement is expected. Based on the above results from two field sites, Carsel et al. [4], concluded that "using the best estimates of transport and transformation properties of metalaxyl and limited calibration for water balance," the PRZM model was "effective in simulating the important processes operating on the pesticide" under field conditions.

Closure

The SESOIL testing study by Watson and Brown [2] is one of a number of such model testing studies that have been performed. We have discussed it here because it is a recent comprehensive study and includes comparisons of field data with predictions by the three models discussed in this manual. Carsel et al. [4], is a good example of the problems encountered in model performance testing using environmental fate monitoring data collected as a part of

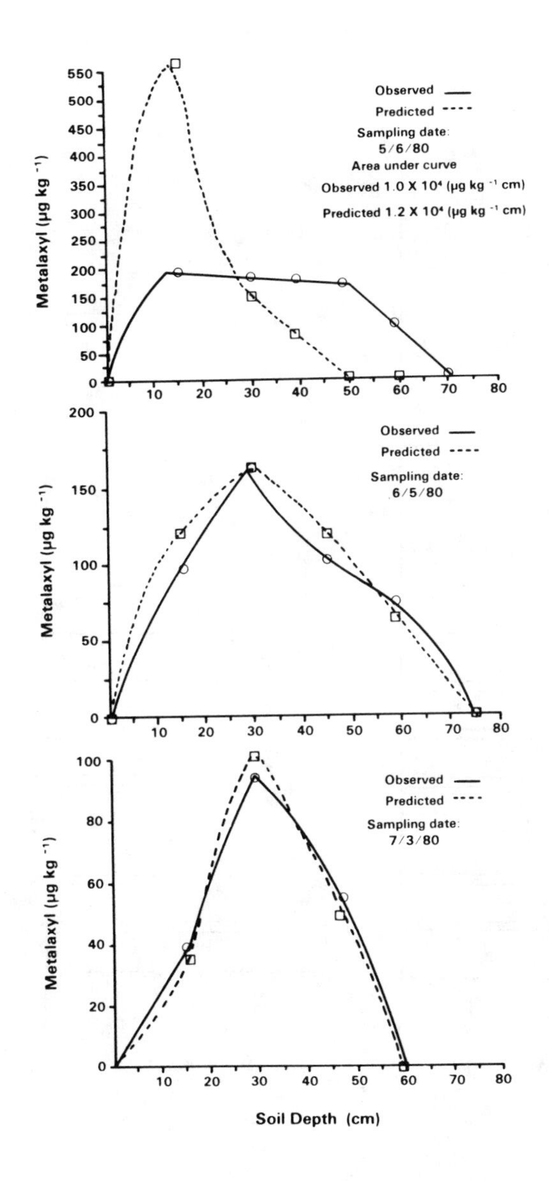

Figure 5-7. Comparison of measured metalaxyl concentration profiles at three sampling dates with those predicted by PRZM model (adapted from Carsel et al. [4]).

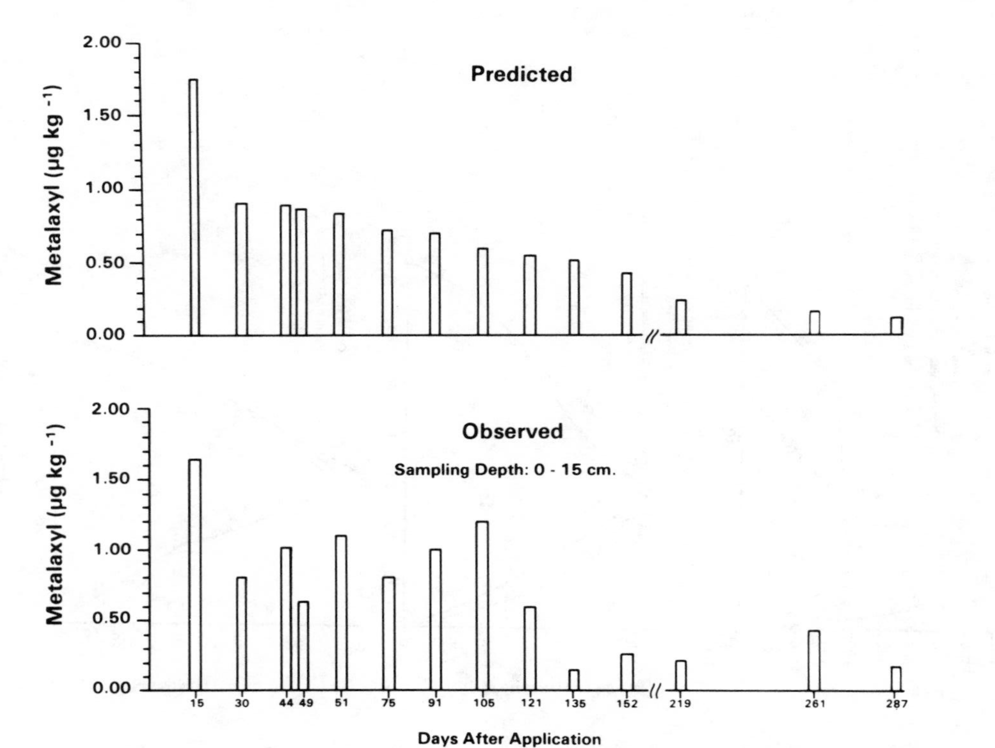

Figure 5-8. Comparison of measured and predicted metalaxyl concentrations in the 0-15 cm depth increment. (Reprinted with permission from Environ. Toxicol. Chem. 5:345-353, R. F. Carsel, W. B. Nixon, and L. G. Balentine, "Comparison of Pesticide Root Zone Model Predictions with Observed Concentrations for the Tobacco Pesticide Metalaxyl in Unsaturated Zone Soils," Copyright, 1986, Pergamon Press, Ltd.).

the pesticide registration process and is not specifically for model evaluation as an objective.

Chapter 1 noted a number of additional model testing studies that provide valuable information and examples of model testing procedures and sensitivity behavior of the specific models discussed herein. In addition to the specific studies on SESOIL and PRZM noted above, the reader should review some of the following studies for the particular model being applied:

a. PESTAN has been applied by Jones and Back [33] to the aldicarb data in Florida, and it has been used by the EPA Office of Pesticide Programs for a variety of compounds in different regions [34].

b. SESOIL has been applied to leaching of selected metals in Kansas and two organics in Montana [35], has undergone sensitivity analyses for different climates and soil types [36], has been tested for hydrologic prediction accuracy [37], and has been compared to a variety of other models for simulating vertical flow in a landfill [38].

c. PRZM has been tested against data on aldicarb in New York, and against atrazine and chloride applied to corn in Georgia [39].

The model validation and testing process has been discussed by Donigian [40] in terms of the most likely reasons for a difference between model predictions and field observations. He identified and discussed four categories of potential errors or discrepancies that often occur, including input data, parameter values (including calibration), model representation, and output (or observed data). Based on the discussion by Donigian [40] and the SESOIL testing described above, the following recommendations are provided to potential model users.

a. Be sure to make conditions under which the model operates (e.g., input meteorologic data, parameter values) as close as possible to actual field conditions under which observed data were collected.

b. Be aware of model assumptions and limitations with respect to representing field conditions (e.g., the inability of SESOIL to accurately represent low permeability soil layers). Biased parameter values can result through calibration if model limitations are not respected.

c. Be especially wary of calibration efforts that result in parameter values outside their normal range of expected values (e.g., K_{OC} values for aldicarb in the SESOIL testing) unless these values are justified by local conditions.

d. Be sure that the model is operating correctly both in terms of algorithm calculations, and proper usage and interpretation of input values. Simple hand calculations to confirm decay rates and to check mass conservation should be performed.

e. Be aware of possible errors, omissions, or inaccuracies in the observed data especially with regard to their possible impact on the calibration process (e.g., calibrating to inaccurate data). The problem of spatial variation must be considered.

In summary, model users should develop an attitude of "informed skepticism" when performing model testing. The user must be fully knowledgeable and informed of model assumptions and limitations and of field conditions; this must be balanced by a healthy skepticism or by a questioning approach, in the interpretation of both the model predictions and the observed data. In this way, models can be used appropriately as tools to aid environmental decision-makers but cannot be used as a crutch to support and defend pre-conceived policy and regulations.

REFERENCES

1. Wagenet, R. J. and P. S. C. Rao. "Basic Concepts of Modeling Pesticide Fate in the Crop Root Zone," Weed Science 33 (Suppl. 2): 25-32 (1984).

2. Watson, D. B. and S. M. Brown. "Testing and Evaluation of the SESOIL Model" (Palo Alto, CA: Anderson-Nichols Co., Inc., 1984).

3. Jones, R. L., P. S. C. Rao, and A. G. Hornsby. "Fate of Aldicarb in Florida Citrus Soils: 2. Model Evaluation," Proceedings of the NWWA/U.S. EPA Conference on Characterization and Monitoring of the Vadose (Unsaturated) Zone, Las Vegas, NV (1983).

4. Carsel, R. F., W. B. Nixon, and L. G. Balentine. "Comparison of Pesticide Root Zone Model Predictions with Observed Concentrations for the Tobacco Pesticide Metalaxyl in Unsaturated Zone Soils," Environ. Toxicol. Chem. 5:345-353 (1986).

5. van Genuchten, M. Th. and W. J. Alves. "Analytical Solutions of the One-Dimensional Convective Dispersive Solute Transport Equation," Technical Bulletin No. 1661, U.S. Department of Agriculture (1982).

6. Bonazountas, M., and J. Wagner. "Pollutant Transport in Soils via SESOIL," Presented at ASCE National Conference of Environmental Engineering, Minneapolis, MN, July 15-16, 1982.

7. Bonazountas, M., and J. Wagner. "SESOIL: A Seasonal Soil Compartment Model" (Cambridge, MA: Arthur D. Little, Inc., 1984).

8. "Testing for Field Applicability of Chemical Exposure Models," Results of the Workshop on Field Applicability Testing, U.S. Environmental Protection Agency, Athens, GA (1982).

9. Standard Practice for Evaluating Environmental Fate Models of Chemicals (Philadelphia, PA: American Society for Testing and Materials, 1984).

10. Hornsby, A. G., P. S. C. Rao, W. B. Wheeler, P. Nkedi-Kizza, and R. L. Jones. "Fate of Aldicarb in Florida Citrus Soils: 1. Field and Laboratory Studies," Proceedings of the NWWA/ U.S. EPA Conference on Characterization and Monitoring of the Vadose (Unsaturated) Zone, Las Vegas, NV (1983).

11. INTERA Environmental Consultants, Inc. "Mathematical Simulation of Aldicarb Behavior on Long Island: Unsaturated Flow and Ground-Water Transport" (1980).

12. Smith, C. N., R. A. Leonard, G. W. Langdale, and G. W. Bailey. "Transport of Agricultural Chemicals from Small Upland Piedmont Watersheds," EPA-600/3-78-056, IAG No. IAG-D6-0381 (Athens, GA: U.S. EPA and Watkinsville, GA: USDA, 1978).

13. Johnson, H. P. and J. L. Baker. "Field-to-Stream Transport of Agricultural Chemicals and Sediment in an Iowa Watershed: Part I. Data Base for Model Testing (1976-1978)," EPA-600/ 3-82-032 (Athens, GA: U.S. EPA, 1982).

14. Wyman, J. A., J. O. Jensen, D. Curwen, R. L. Jones, and T. E. Marquardt. "Effects of Application Procedures and Irrigation on Degradation and Movement of Aldicarb Residues in Soil," Submitted for publication (1984).

15. Jury, W. A., L. A. Stolzy, and P. Shouse. "A Field Test of the Transfer Function Model for Predicting Solute Transport," Water Resour. Res. 18:369-375 (1982).

16. Jones, R. L., and G. W. Gee. "Assessment of Unsaturated Zone Transport for Shallow Land Burial of Radioactive Waste: Summary Report of Technology Needs, Model Verification, and Measurement Efforts (FY78-FY83)," PNL-4747 (Richland, WA: Battelle, Pacific Northwest Laboratory, 1984).

17. Rao, P. S. C., J. M. Davidson, and L. C. Hammond. "Estimation of Non-Reactive and Reactive Solute Front Locations in Soils," EPA-600/9-76-015, Proceedings of the Hazardous Wastes Research Symposium, Tucson, AZ (1976).

18. Enfield, C. G., R. F. Carsel, S. Z. Cohen, T. Phan, and D. M. Walters. "Approximating Pollutant Transport to Ground Water," Ground Water 20:711-722 (1982).

19. Carsel, R. F., C. N. Smith, L. A. Mulkey, J. D. Dean, and P. Jowise. "User's Manual for the Pesticide Root Zone Model (PRZM): Release 1," EPA-600/3-84-109 (Athens, GA: U.S. EPA, 1984).

20. Moore, G. E., T. E. Stoeckenius, and D. A. Stewart. "A Survey of Statistical Measures of Model Performance and Accuracy for Several Air Quality Models," EPA-450/4-83-001 (Research Triangle Park, NC: U.S. EPA, 1982).

21. Ambrose, Jr., R. B., and S. E. Roesch. "Dynamic Estuary Model Performance," Journal of the Environmental Engineering Division, ASCE, Vol. 108, EE1 (1982).

22. Halfon, E. "Regression method in ecotoxicology: A Better Formulation using the Geometric Mean Functional Regression," Environ. Sci. Tech. 19:747-749 (1985).

23. Lyman, W. J., W. F. Reehl, and D. H. Rosenblatt. Handbook of Chemical Property Estimation Methods (New York: McGraw-Hill Book Company, 1982).

24. Hough, A., I. J. Thomason, and W. J. Farmer. "Behavior of Aldicarb in Soil Relative to the Control of Heterodera schachtii," J. of Nematology 7:214-221 (1975).

25. Bromilow, R. H., R. J. Baker, M. A. H. Freeman, and K. Gorog. "The Degradation of Aldicarb and Oxamyl in Soil," Pesticide Sci. 11:371-378 (1980).

26. Supak, J. R. "The Volatilization, Degradation, Adsorption and Desorption Characteristics of Aldicarb in Soils and Clays." Ph.D. Dissertation, Texas A & M University (1972).

27. Smelt, J. H., M. Leistra, N. W. H. Houx, and A. Dekker. "Conversion Rates of Aldicarb and its Oxidation Products in Soils. I. Aldicarb Sulfone," Pesticide Sci. 9:293-300 (1978).

28. Smelt, J. H., M. Leistra, N. W. H. Houx, and A. Dekker. "Conversion Rates of Aldicarb and its Oxidation Products in Soils. II. Aldicarb Sulfoxide," Pesticide Sci. 9:286-292 (1978).

29. Webster, R. and T. M. Burgess. "Sampling and Bulking Strategies for Estimating Soil Properties in Small Regions," J. Soil Sci. 35:127-140 (1984).

30. Webster, R. and S. Nortcliff. "Improved Estimation of Micronutrients in Hectare Plots of the Soning Series," J. Soil Sci. 35:667-672 (1984).

31. Khan, M. A., and S. Nortcliff. "Variability of Selected Soil Micro-Nutrients in a Single Soil Series in Berkshire, England," J. Soil Sci. 33:763-770 (1982).

32. Carsel, R. Personal Communication (1986).

33. Jones, R. L., and R. C. Back. "Monitoring Aldicarb Residues in Florida Soil and Water," Environ. Toxicol. Chem. 3:9-20 (1984).

34. Lorber, M. Personal Communication (1985).

35. Bonazountas, M., J. Wagner, and B. Goodwin. "Evaluation of Seasonal Soil/Groundwater Pollutant Pathways" (Cambridge, MA: Arthur D. Little, Inc., 1982).

36. Wagner, J., M. Bonazountas, D. M. Alsterberg. "Potential Fate of Buried Halogenated Solvents via SESOIL" (Cambridge, MA: Arthur d. Little, Inc., 1983).

37. Hetrick, D. M. "Simulation of the Hydrologic Cycle for Watersheds," Proceedings for the Applied Simulation and Modeling Conference, San Francisco, CA (1984).

38. Kincaid, C. T., J. R. Morery, S. B. Yabusaki, A. R. Felmy, and J. E. Rogers. "Geohydrochemical Models for Solute Migration, Volume 2, Preliminary Evaluation of Selected Computer Codes for Modeling Aqueous Solutions and Solute Migration in Soils and Geologic Media," EA-3417 (Palo Alto, CA: Electric Power Research Institute, 1984).

39 Carsel, R. F., L. A. Mulkey, M. N. Lorber, and L. B. Baskin. "The Pesticide Root Zone Model (PRZM): A Procedure for Evaluating Pesticide Leaching Threats to Groundwater," Ecol. Model. 30:49-69 (1985).

40. Donigian, Jr., A. S. "Model Predictions vs. Field Observations: The Model Validation/Testing Process," in Fate of Chemicals in the Environment, R. L. Swann and A. Eschenroeder, Eds., ACS Symposium Series 225 (Washington, DC: American Chemical Society, 1983).

SECTION II

CHEMICAL PROCESSES, PARAMETER ESTIMATION, AND VARIABILITY IN THE VADOSE ZONE

CHAPTER 6

CHEMICAL MOVEMENT THROUGH SOIL

W. A. Jury
University of California-Riverside, Riverside, California

PROCESS DESCRIPTION

Transport Mechanisms

Chemicals are transported through soil principally by three mechanisms: mass flow of dissolved chemical within moving soil solution, liquid diffusion within soil solution, and gaseous diffusion within soil air-filled voids. The first mechanism, mass flow, refers to the passive transport of dissolved solute within moving soil water. To a first approximation, the solute within an arbitrary volume element of moving soil water is assumed to be uniformly distributed, and hence the mass flux of chemical is given by the product of the volume flux of water times the dissolved solute concentration expressed in units of mass of solute per volume of water (see Appendix A). Liquid diffusion refers to the transport of the dissolved solutes by diffusion in response to molecular scale collisions. The effect of large numbers of these collisions is to move dissolved solutes from regions of higher solute density to lower solute density; thus, the diffusion flux is proportional to the density or concentration gradient. The coefficient of proportionality between the diffusion flux and the concentration gradient is called the liquid diffusion coefficient (see Appendix A). Chemical vapor molecules within the soil air spaces also undergo molecular collisions and spread out by vapor diffusion. As in the case of liquid diffusion, the flux of vapor molecules is proportional to the density or concentration gradient. The coefficient of proportionality is called the vapor diffusion coefficient (see Appendix A).

An additional term is frequently required in the mathematical description of chemical transport. Since the soil water flux is represented as a continuous quantity which is volume-averaged over many pores, the individual complicated water flow paths around soil grains are mathematically replaced by an equivalent one-dimensional flow. When this one-dimensional flow of water is multiplied by the dissolved solute concentration, the resulting solute mass flux does not take into account the additional spreading of solute which occurs by three-dimensional mass flow at the pore scale in the actual system but which is not represented in the volume-averaged mathematical treatment. This apparent solute diffusion arising from the mass flux effects which are obscured by mathematical volume averaging is called hydrodynamic dispersion [1]. Under certain conditions, this dispersion transport process is mathematically equivalent to transport by liquid diffusion and may be included in the transport equations (see Appendix A) by using an effective liquid diffusion-dispersion coefficient to account for spreading of liquid solute molecules. Because this effect depends on the size of the volume averaging, it is a function of the scale of approximation of the water flux. Thus, hydrodynamic dispersion is considerably more important on the field scale than on the laboratory scale [1,2].

Soil, Environmental, and Management Factors Influencing Chemical Transport Through Soil

There are a host of factors influencing each of the transport mechanisms mentioned above. In the mathematical derivation of the transport equations and the boundary conditions given in Appendix A, the influence of many of these parameters is expressed through the mathematical relationships. In this discussion, reference to these mathematical equations will be made where appropriate; however, most of the discussion will be qualitative, emphasizing the physical rather than mathematical connection between a parameter and a transport mechanism.

Table 6-1 summarizes the various parameters influencing chemical transport through soil and is somewhat arbitrarily divided into groups of soil parameters, environmental parameters, and management parameters. In the discussion below, the effect of each of these parameters on the four transport mechanisms of mass flow, liquid diffusion, liquid dispersion, and gas diffusion will be discussed; literature information will be brought in where appropriate.

Soil Parameters

Soil water content. Volumetric soil water content (θ) has a significant influence on the liquid and gaseous diffusion transport mechanisms. The mass of chemical moved per unit time from point A to point B by diffusion is inversely proportional to the distance between A and B [3]. Thus, the actual path length in soil followed by a vapor molecule or a dissolved solute molecule is strongly affected by water content. For liquid diffusion, diffusive transport increases as water content increases because the cross-sectional area available for flow increases, and the path length decreases as liquid replaces air in the medium. Conversely, for vapor diffusion, the transport by diffusion decreases with increasing water

Table 6-1. Soil, Environmental, and Management Parameters Influencing Chemical Transport Through Soil.

Soil Parameters	Environmental Parameters	Management Parameters
Water Content	Temperature	Chemical Concentration
Bulk Density or Porosity	Precipitation	Irrigation Management
Permeability (Saturated)	Evapotranspiration	Crop Characteristics
Clay Content		
Surface Area		
Organic Matter Content		
Depth to Ground Water		
Water Retention (Field Capacity)		

content because the air space decreases. In fact, as shown in Appendix A, the volumetric air content (a) which is equal to the porosity (Ø) minus the volumetric water content (θ) plays the same functional role in vapor diffusion as the liquid water content plays in liquid diffusion (see equations A-9 and A-11). Models which describe the change in liquid or vapor diffusion coefficient as a function of water content are called tortuosity models [3]. Although there are a number of different kinds of tortuosity models which have been proposed over the years, the most versatile appears to be the Millington and Quirk model discussed in Appendix A [4]. In this model, the liquid diffusion flux is proportional to the 10/3 power of the water content, and the vapor diffusion flux is proportional to the 10/3 power of the air content. This means that the amount of solute transported by each mechanism will change dramatically as water content changes [5].

Since mass flux of chemical is proportional to water flux multiplied by dissolved solute concentration, it is not directly affected by water content. Since increasing the water content of a given soil will result in higher mass flux, some correlation between chemical movement and water content may be found. Hydrodynamic dispersion is also proportional to water flux. Thus, it is usually not considered to be a function of water content.

The partitioning of chemicals between gaseous, liquid, and solid phases can obviously affect chemical transport by affecting the amount of chemical in solution and in the gaseous phase. When soil water contents decrease below a few monolayers of water (which can occur in the top few millimeters of soil during intense drying) water molecules which preferentially occupy soil adsorption sites are displaced from the surface, and the chemical adsorption capacity of the soil is greatly increased [6]. This increased adsorption capacity causes liquid and gaseous concentrations to decrease dramatically in this dry region, and all four of the above transport mechanisms are reduced significantly as most of the chemical is displaced to adsorption sites. However, this adsorption effect does appear to be reversible when the soil surface layer rewets [6,7]. For this reason, it may be possible to ignore this aspect of the water content dependence of adsorption provided that the time period during which the surface layer is dry is short-lived. Should a permanently dry surface layer (e.g., a desert soil) be part of a scenario, the effect should definitely be considered.

Bulk density or porosity. Soil porosity ($\emptyset$) is related to bulk density (ρ_b) by a linear relation

$$\emptyset = 1 - \rho_b/\rho_m \qquad (6\text{-}1)$$

where ρ_m is soil mineral density (g/cm^3) which for most soils lies between 2.65 g/cm^3 (clays) and 2.75 g/cm^3 (sands). For this reason, it suffices to discuss only the dependence of the soil porosity ($\emptyset$) on chemical transport with the understanding that decreased porosity implies increased bulk density.

Decreasing soil porosity or increasing bulk density will generally decrease chemical transport by each of the above mechanisms. In the case of liquid or vapor diffusion, decreasing porosity generally decreases the cross-sectional area available for flow and increases the path length by placing more solid obstacles per unit volume in the way of the diffusing molecules.

Although the equations in Appendix A specifically contain a factor of $\emptyset^2$ in the denominator of the tortuosity factor, this implied functional dependence is misleading because the volumetric air and water contents themselves depend on porosity. It is perhaps most instructive to examine the two equivalent problems: vapor transport in very dry soil or liquid transport in very wet soil. In each case, the tortuosity factor and hence the transport flux according to the Millington and Quirk model is proportional to the 4/3 power of the porosity. Thus as porosity decreases, both liquid and vapor diffusion decrease somewhat faster than linearly.

Transport by mass flow is indirectly affected by porosity since regions of low porosity are likely to have lower permeability to water transport. Although no good structural models exist for the relationships between porosity and permeability, permeability of a given soil type strongly decreases as porosity decreases because the pore sizes contract. However, finer-textured soils such as clays generally have a higher porosity and lower permeability than sandy soils.

Hydrodynamic dispersion has not been quantitatively linked to porosity. However, it is proportional to water flux (see equation A-13) which in turn is decreased by decreases in porosity under most conditions.

Another influence of porosity on transport which equally affects all mechanisms is that decreasing soil porosity increases the density of mineral and organic adsorption sites and thus causes increased adsorption of chemical with a corresponding decrease in solution and gaseous concentration. Since transport only occurs in liquid or vapor phases, transport by all four mechanisms decreases as porosity decreases.

Saturated hydraulic conductivity or permeability. Saturated hydraulic conductivity (K_S) is the proportionality coefficient between the saturated water flux and the hydraulic head gradient [8]. An equivalent index frequently used by engineers is the intrinsic permeability (k) which is equal to $K_S \gamma / \rho g$ where γ is the viscosity of water, ρ is water density, and g is the gravitational acceleration. The permeability k thus defined is independent of any fluid properties. For any process where water is ponded on the soil surface, either through intense irrigation or rainfall or because of a deliberate ponding management such as a waste holding pond, permeability will have a dominant influence on the amount of water infiltrating into the soil and hence will strongly influence mass flow and dispersion. When water flow occurs at an unsaturated water content (such as when water is applied at the soil surface at a constant rate less than the saturated hydraulic conductivity), the main influence of the soil hydraulic conductivity is to regulate at what water content the flow will occur. There is no relationship between hydraulic conductivity and liquid or vapor diffusive transport.

Clay content. The clay content of soil is a qualitative index which has been correlated with a number of other soil properties such as water-holding capacity, specific surface area, etc. One of the highest correlations is between clay content and saturated hydraulic conductivity or permeability since soils high in clay tend to be low in permeability [9]. Thus, all of the arguments made above about permeability also apply to clay content. Clay content is also highly correlated with ion exchange capacity and therefore will strongly influence the adsorption of chemical ions. For most non-polar organic chemicals, however, there is usually no correlation between clay content and adsorption [10]. There is also a reasonably high negative correlation between clay content and soil porosity. Therefore, the comments made above about porosity also apply to clay content.

Adsorption site density. The density of adsorption sites for chemicals may be represented by two other soil indices which are specific surface area (surface area per soil volume or mass) and soil organic matter content.

Since clays have a large specific surface area, the comments made above about clay content also apply to surface area. Thus, the correlation between surface area and organic chemical adsorption is likely to be quite low. However, organic matter content has been found to be positively correlated with organic chemical adsorption in a number of studies [10,11]. Thus, one immediate effect of increasing organic matter content is to increase the extent of chemical adsorption and thus decrease liquid and gaseous concentration thereby strongly decreasing the extent of transport by each of the four transport mechanisms.

Depth to ground water. Depth to ground water can strongly influence the extent of upward water flow occurring to a surface layer which has been evaporating without water input for an extended period of time. It has been shown both theoretically and experimentally [12] that finer textured soils can move water (and hence chemicals by mass flow) upward from much greater depths than can coarse textured soils. For this reason, salinization of a surface soil layer from a saline water table is an ever-present problem above shallow saline water tables such as near the Salton Sea area of the Imperial Valley of California.

Depth to ground water also obviously affects the travel time of a chemical leached from the surface to ground water. However, depth to ground water has no direct influence on downward transport processes of any of the four mechanisms other than to create a relatively thin region known as the capillary fringe which is above the ground-water table and which has a high water content.

Water retention (field capacity). An extremely important property of soil which can influence the above processes particularly when inputs of water to the soil surface are infrequent is the amount of water remaining after drainage has become insignificant. This water content, known as the field capacity, is much higher in a finer textured soil than in a sandy soil [8]. Its influence on chemical transport is identical to that of soil water content above.

Environmental Parameters

Temperature. Temperature affects all rate processes in nature including diffusive transport. The diffusion coefficient of a chemical moving through pure water is an increasing function of temperature as is the gaseous diffusion coefficient of a chemical moving through pure air. In addition, chemical vapor pressure is a strongly increasing function of temperature, and hence as temperature increases, the fraction of the total chemical present in the vapor phase increases (see equation A-17). For this reason, vapor diffusion is much more strongly enhanced by increases in temperature than is liquid diffusion. However, transport by both mechanisms will increase for the reasons given above. There have been various models for describing the temperature dependence of the vapor and liquid diffusion coefficients. Standard chemical engineering textbooks indicate that vapor diffusion coefficients have a temperature dependence proportional to $T^{1.75}$ [13]. They also indicate that liquid diffusion coefficients increase with temperature, but no standard functional relationship applies.

The temperature dependence of chemical mass flow depends on the temperature dependence of water flow for reasons given above. Liquid water transport under temperature gradients is poorly understood at present and is likely to be a relatively minor effect compared to the other factors influencing water transport [14]. Thus, the primary factor affecting the temperature dependence of chemical mass flow is the temperature dependence of the liquid concentration in a three-phase soil system. The amount of chemical present in the liquid phase is strongly affected by absorption which generally decreases as temperature increases [15]. Thus, increased liquid concentrations resulting in increased mass flows would be expected as temperature increases.

Precipitation. The characteristics of the precipitation or rainfall events (i.e., intensity and distribution) will have a critical influence on the extent of chemical transport in all phases above. Precipitation will have a dominant influence on mass flow because the rainfall rate is directly related to the water flow rate in the soil. Thus soils with intense, frequent rainfall will have high water fluxes and hence high chemical mass fluxes and dispersion fluxes. Furthermore, an extremely intense rainfall event might induce saturation which could result in a greatly enhanced mass transport through soils of high permeability. The exact response of a soil to a rainfall event can be determined only by solving the transport equations given in Appendix A. However, it is reasonable to conclude that for a given soil the effect of increasing the rainfall intensity in a given time period will be to increase the mass flux and also to increase the water content. The effect of the latter will be to increase the extent of transport by liquid diffusion and to decrease the extent of transport by vapor diffusion.

Evapotranspiration. Evapotranspiration represents the amount of applied water which is removed by plants or water loss from surfaces and hence is unavailable for drainage. Thus the extent of evapotranspiration will strongly affect the water flux below the root zone and hence the extent of chemical leaching by mass flow. For soils not receiving water input by irrigation, rainfall minus evapotranspiration determines the net amount of water leaching beyond the crop root zone. This parameter, if positive, will imply long-term drainage or downward movement, and, if negative, will imply drying processes which will lower water content and induce upward flow from ground water.

Water loss by evapotranspiration (ET) may be regulated by external meteorological conditions (potential ET) or by resistances in the soil or plant. The latter generally dominates only when the soil is dry. The most important meteorological conditions regulating potential ET are solar radiation, wind, air temperature, and air humidity. These meteorological measurements may be used in formulas such as the Penman combination equation to predict potential ET for a crop or soil surface [16].

Management Parameters

Chemical concentration. To the extent that the amount of chemical applied to soil is a management option, for example with pesticide application, this variable has been characterized as a management parameter. The effect of increasing chemical concentration on transport is always to increase the amount of transport in each phase by increasing the concentration in each phase. In fact, for systems represented by linear partition coefficients (see equations A-27 through A-30), each of the four transport mechanisms are proportional to total chemical concentration because a unit increase in chemical concentration partitions proportionally into each phase. However, not all soils adsorb solute linearly. In many cases, chemicals applied at higher concentrations are less efficiently adsorbed than are chemicals applied at low concentrations. In such cases, the amount of mass flow may increase more than linearly with increases in chemical concentrations.

Irrigation management. Application of irrigation water to soil has an effect similar to the rainfall application. However, since irrigation can be more carefully controlled than rainfall, it is characterized as a management parameter. For salinity control, crops must be irrigated at a sufficient volume over the evapotranspiration rate in order to produce drainage. As a practical matter, it is rarely possible to manage irrigation with less than a leaching fraction (drainage divided by irrigation) of 0.2. While chemicals are moving through a crop root zone, chemical concentrations are increased as water is preferentially extracted and solute left behind by plant roots. Thus, one effect of irrigation management is indirectly to raise chemical concentrations in the root zone. However, at the same time, irrigation management can directly control mass fluxes by restricting the amount of salt added with irrigation water and by decreasing drainage flux. Thus, it has a complicated effect on mass transport. The other main influence of irrigation management on transport is to affect the water content of the crop root zone. Irrigating at a higher intensity will increase water content and thus affect the diffusive transport mechanisms as indicated in the earlier discussion.

Crop characteristics. General crop characteristics such as wilting point and rooting depth will indirectly affect chemical transport. The major effect of wilting point is to dramatically decrease evapotranspiration when water application to the crop either by rainfall or irrigation becomes inadequate. Rooting depth can somewhat affect the transport mechanisms by widening or narrowing the zone of extraction. However, since the total amount of water extracted is the same in each case, this effect is relatively minor and generally not given much importance in chemical transport.

Crop residues may influence chemical transport by increasing organic matter in the soil, with the result that the number of adsorption sites will also be increased. Also, channels left by decaying plant roots may induce preferential flow of water and chemical at high rates through these macropores.

Chemical and water application variability. When one-dimensional solute transport models are used to describe downward movement below fields which have received a common surface application of water and chemical, part of the variability of observed solute concentrations is caused by non-uniform application of both water and solute. Water application methods such as furrow or flood irrigation which pond water over the surface are highly variable and commonly result in coefficients of uniformity (CU) of 50 percent or lower whereas sprinkler or drip systems applied during periods of low wind may approach CUs of 90 percent [17].

Chemical application variability is also quite high particularly when the chemical is spread or sprayed on the surface rather than added with irrigation water [18]. Taylor and Klepper [19] observed a CU of only 20 percent in dieldrin concentrations sprayed onto a 0.13 ha plot. Similarly, Richter [20] observed CUs of 50 percent on two fields sprayed with potassium bromide. Finally, El Abd [21] found a CU of 50 percent in recovered napropamide sprayed on a 1.4 ha field. These large spatial variations must be taken into account in sampling strategies for model validation.

MATHEMATICAL DESCRIPTION OF CHEMICAL MOVEMENT

The mathematical differential equations describing vertical flow of water and chemicals in soil are derived in Appendix A. In this Appendix, it is shown that the equations result from an application of conservation of mass (either water or chemical) within a unit volume of soil together with an appropriate expression for the flow of material per unit area per unit time using recognized transport laws.

Appendix A also discusses the appropriate boundary conditions for use in solving chemical and water transport processes and reviews standard methods of solution of the mathematical equations and their boundary conditions.

LIMITATIONS OF THE PHYSICAL TRANSPORT MODELING APPROACH

Non-Equilibrium Effects

Chemical adsorption or reaction is ultimately a rate-limited process. In soil solution, the approach to equilibrium is a function of, among other factors, soil geometry and water flow. At the present time, no model exists for describing the dependence of the rate coefficient in expressions such as equation (A-26) of Appendix A on these variables. Rao et al. [22], have made a promising first approach in describing this functional dependence for simple spherical aggregates in a soil column, but their method has not been tested under more natural conditions. In the interim, equilibrium models have usually been used particularly under natural field conditions when geometry is poorly characterized. At the present time, such equilibrium models must be used but will be most accurate when the time required to reach equilibrium is short compared to the residence time of a chemical in the soil region of interest.

Measurement Limitations

There are a number of practical limitations to model calibration and testing under field conditions. Table 6-2 lists a number of potential problems encountered in field validation of water and chemical transport models together with the implications of the problem. These will be discussed in more detail below.

Lateral and Vertical Variability of Transport and Retention Properties

In recent years considerable attention has been focused on the spatial variability of physical and chemical properties at the field scale [2,3,23,24,25]. This research, still more basic than applied, has pointed out that in the field, physical and chemical properties have a spatial correlation which persists over short length scales, as little as 5 meters, and that properties associated with transport may have high coefficients of variation over field areas. Chapter 11 of this report provides a review of some major experimental studies of spatial variability.

What these findings indicate is that a serious problem lies ahead in both field characterization and model calibration when large areas are involved. For example, Biggar and Nielsen [2] estimated that for their 150 ha field, hundreds of measurements of the effective chloride dispersion coefficient D_E would have to be taken to obtain the field-wide average value to within 50 percent of the true mean. Such variability has led to new approaches for designing measurement strategies and subsequently modeling chemical transport events [26]. Even on scales smaller than 1 ha fields, however, variability will likely limit the success of model calibration or testing. Because of the innate variability, many replicates will be needed to characterize any of the parameters discussed above and summarized in Table 6-1. Furthermore, functional relations which are required to characterize hydraulic conductivity as a function of water content, for example, are greatly obscured by variability between replicate measurements and functional relations. In principle, each location in the soil could have its own functional relationship, but linking these together would require a

Table 6-2. Potential Problems Encountered in Field Validation of Water and Chemical Transport Models.

Problem	Implications
1. Lateral and vertical variability of transport and retention properties	1. Many sample replicates are needed to characterize mean values. Functional relations may be obscured by variability.
2. Macropores, cracks, plant and animal holes, holes, interspersed with bulk structural characteristics	2. Isolated soil structural features may have significant influence on material transport which is difficult to characterize by measurement.
3. Time dependent boundary conditions and soil response characteristics	3. Oscillations in input create cyclic water and temperature regimes near surface. Hysteresis and temperature effects on transport may influence observations. Large water fluxes during storms may introduce rate-limited adsorption and macropore flow.
4. Crop growth	4. Growth will change rooting density and water uptake patterns, and alter evaporation and transpiration partition at canopy surface.
5. Difficulty in sampling or measuring important transport properties below surface zone	5. Surface-measured properties must be used to represent greater depths, creating possible errors.
6. Inability to measure water drainage flux *in situ*	6. Water and chemical mass fluxes cannot be inferred from point measurements of concentration.
7. Scale effects on transport	7. Field scale transport of chemicals may involve new transport mechanisms not present in lab-scale equations.

complicated three-dimensional model which is not currently in existence.

Structural Effects

In any natural setting, inter-aggregate structural features such as soil macropores, cracks, plant root holes, or animal burrows will be interspersed with the bulk structural characteristics. These geometric features although very small in volume can have a significant effect on the transport of chemicals and particularly on those which are highly adsorbed [27]. For soil containing large cracks and holes, the dominant flow mechanisms for water and bulk chemical flow may not even be those described in the physical transport theory given in Appendix A. Characterization of the influence of these structural voids on transport properties has not been attempted on a field scale although proposals have been made to characterize them in a probabilistic sense [28].

A major problem with such structural voids is that they are very difficult to detect through antecedent soil measurements or calibrations. In the future, it will likely require several detailed field experiments in order to characterize their statistical probability in several representative soil types. In the interim, it must suffice only to lump them with other causes of variability in the field and to use the extent of measurement variability as a guide in designing sample numbers [26].

Time Dependent Boundary Conditions and Soil Response Characteristics

In any natural setting, surface boundary conditions are dominated by the diurnal cycling of radiation at the soil surface. As a response to this diurnal variation, temperature profiles near the surface fluctuate between limits, and the soil profile near the surface undergoes wetting and drying cycles. As mentioned above, both water and chemical transport and retention parameters have been observed to be hysteretic. Therefore, in principle, hysteresis can affect measurements of parameters of interest near the soil surface when conducted under natural conditions. Further, and perhaps more significantly, temperature variations in response to diurnal cycling are likely to be extreme in the top 30 cm of soil. The influence of such temperature variations on transport and retention is only partially understood and greatly increases the requirements of a data base. With the exception of several isolated research studies [29,30], the influence of temperature variations on water transport in the liquid phase is not well understood. Even less well understood is the influence of temperature variations on chemical transport. Thus, virtually all models assume isothermal non-hysteretic conditions in the absence of any theoretical or experimental information available. Nevertheless, these effects may influence observations made near the soil surface.

Time-dependent water inputs due to intense storms may also result in large infiltration fluxes near the soil surface. In addition to enhancing transport by mass flow and dispersion of chemicals, such large water fluxes may increase the influence of structural voids discussed above and may also enhance the rate-limited non-equilibrium adsorption by decreasing the residence time in the soil compartments near the soil surface. As mentioned above, none of these responses to the effects of water flow variations are taken into account in most models.

Crop Growth

Models of water and chemical movement through cropped fields require experimental values of various crop growth parameters which are difficult to obtain. Particularly difficult to measure is the rooting depth and root density which have critical influences on water uptake patterns. During root growth, these density patterns may change dramatically over short periods of time [19], and unless prohibitive numbers of replicated measurements are made, such time variations may not be characterized by measurement. Similarly, during crop growth, the extent of soil cover at the surface is increasing, and the partitioning between evaporation and crop transpiration is continually changing. Modeling of this changing partition may be required for certain transport scenarios.

Measurement Difficulties at Great Depths

In any practical field experiment, the bulk of soil measurements is taken near the soil surface because of increased sampling difficulties at greater depths. As a consequence, many important transport properties and much structural information about the soil are inadequately characterized at depths below the soil surface. Nonetheless, many transport models and particularly those involving hazardous chemicals are used to project chemical movement all the way to ground water. In such cases, the data base measurements obtained for calibrating and validating such models will necessarily be concentrated near the soil surface, and the properties of the soil below such surface measurements will be obtained by inference or extrapolation. This procedure could have a critical effect on transport predictions far below the soil surface particularly when a large horizon shift or bedrock layer is encountered.

Inability to Measure Water Drainage Flux

Unless a drainage lysimeter is installed, water drainage flux is not measurable in the field. This greatly decreases the amount of validation and testing that can be conducted on any chemical transport model since it prohibits measuring chemical mass flux or water mass flux at various locations in the field. Limiting such estimates to field-wide averages conducted over long time intervals is likely to provide substantially less information about model validity.

Spatial Variability and Volume Averaging

Recent research on leaching of non-adsorbed solutes at the field scale has led to the tentative conclusion that the classical one-dimensional convection-dispersion equation (A-35) which uses a field-wide average downward leaching flux may not be applicable near the soil surface when substantial downward movement due to mass flow is involved [31]. As a result, either a multi-dimensional form of equation (A-35) may have to be used, or some sort of new representation based on a statistical model may have to be attempted. The implications for this and chemical leaching studies are largely unexplored at this time. However, Jury et al. [32], conducted a field trial over the top 3.5 meters of soil using a stochastic-convective model and showed that it gave a superior description of observed chemical leaching to the conventional convection-dispersion equation above. Current approaches to solute transport at the field scale have been recently reviewed in an article by Jury [33].

SAMPLING METHODOLOGY

The parameters listed in Table 6-1 and discussed in the text above should ideally be measured in an undisturbed manner in the actual field setting where the model will be applied. In most cases, this is not possible, and the values must be inferred either from disturbed samples taken back to the laboratory or perhaps even indirectly from behavior of other parameters which is interpreted through a model. Measurement methods currently in use for the major parameters and their limitations will be discussed below.

Static Soil Properties

The easiest of the parameters to measure is the static soil properties which do not vary as a function of time, e.g., water content, or temperature at a given location. For these properties it suffices to take a number of soil cores at representative locations in the field and analyze them in the laboratory. Among the parameters which can be measured this way are: particle size distribution or percent clay which is usually determined by a hydrometer or pipette method [34]; specific surface area of soil which may be determined by a variety of methods including adsorption of ethylene glycol monoethyl ether [35]; and organic carbon fraction which may be determined by CO_2 detection following combustion [36]. In addition, soil bulk density or porosity may be determined by measuring the mass of dry soil in a known volume of a core sample [37]. The major requirement in this measurement is making sure that the soil core carves out an undisturbed volume without compression of the soil inside. For this reason, measurements of porosity or bulk density are usually in error when taken by compression core measurements (such as pounding) deeper than a few meters below the surface.

Saturated hydraulic conductivity may be measured either by steady state ponding on a large surface plot or by transient techniques such as the auger hole or piezometer method [38]. These methods each take an average saturated conductivity over a large volume and in addition display significant variability across the field. For example, Nielsen et al. [39], observed a coefficient of variation of 133 percent in saturated hydraulic conductivity measurements taken over a 150 ha field. They estimated that several hundred measurements would have to be taken to estimate the field mean saturated hydraulic conductivity within 20 percent.

Transport and Retention Functions

Water and chemical transport coefficients which are functions of water content or matric potential display significant variability in a field setting. Even more importantly, however, it has been repeatedly shown that laboratory measurements taken on either disturbed or undisturbed soil cores give values different from measurements taken on the soil profile itself. Therefore, field measurement of these transport parameters is essential. Hydraulic conductivity as a function of water content or matric potential may be obtained dynamically by the instantaneous profile method which measures water contents and matric potentials during drainage following saturation [40]. However, recent investigations have shown that useful information about hydraulic conductivity as a function of water content may be obtained by using the far simpler unit gradient method for measuring hydraulic conductivity [41]. Matric potential (h) as a function of water content (θ) may be measured by measuring the water content near a tensiometer reading; this could be done by use of a neutron probe or by taking soil cores. Each identifiable soil horizon seems to possess a different matric potential water content relationship, $h(\theta)$, so that measurements on depth increments of 30 or more centimeters often give different functional relationships between h and θ even at a single plot. Furthermore, variation among replicates at a given depth across the field is substantial [39].

Hydrodynamic dispersion coefficients do not appear to be strongly dependent on water content but are functions of the average water flux. The dispersivity (see equation A-13) can be measured only by observing solute transport and interpreting the measurements with the convection-dispersion equation (A-35). Since dispersion arises from volume averaging, the extent of dispersion depends on the size of the region being simulated. Therefore, dispersion coefficients are much smaller on the laboratory scale than on the field scale. At the present time there appears to be no *a priori* way of estimating the dispersion coefficient, and it must be measured *in situ* by observing chemical movement on the field scale.

Tortuosity Functions

The tortuosity functions which express the reduction in diffusion due to soil (see Appendix A) have been fitted to many different models over the years (see review by Sallam et al. [42]). However, few or no measurements of these functions have taken place in field conditions because the volumetric water content may vary in three dimensions. Therefore, tortuosity is unlikely to be measurable in a natural setting and is likely to be replaced by an assumed model form. At the present time the Millington and Quirk model (equation A-9) appears to be the most versatile function for representing the effects of water content on gas or liquid diffusion in soil.

Chemical Properties

The Henry's constant K_H (equation A-22) describing the ratio of gaseous and liquid concentrations at equilibrium is assumed to be the same in soil as it is in air. For organic chemicals such as pesticides, Spencer and Cliath [43] have shown that for several chemicals, Henry's Law of proportionality holds all the way up to saturation chemical concentrations. Therefore, for these chemicals at least, the Henry's constant may be set equal to the ratio between saturated vapor density and solubility. The major problem with the estimation of Henry's constant at the present time appears to be agreement on a standard protocol for measuring vapor pressure and solubility. As pointed out by Spencer and Cliath [44], improvement of protocol has led to a decrease in the error among Henry's Law determinations for a single chemical by different investigators.

Chemical Adsorption Coefficient

At the present time, most chemical adsorption coefficient measurements are made by batch equilibrium shaking of completely dispersed soil samples allowed to come to equilibrium in the laboratory. The resulting functional relationship between liquid concentration and adsorbed concentration is fitted to an adsorption model such as the Freundlich relationship (equation A-24). However, recent work has shown that adsorption under natural conditions on structured soils is likely to be significantly less than that estimated from the batch equilibrium measurement because the adsorption surfaces are incompletely exposed during dynamic transport. For this reason, column flow-through methods have also been used to measure adsorption behavior [45], and these methods do not always measure the same values as the batch methods. For example, Jury et al. [46], showed that pesticide K_D values at 36 sites on a field soil obtained by batch equilibrium methods were different from those obtained by observing leaching patterns of undisturbed soil columns taken from the same 36 sites. In addition, each measurement showed substantial variability between replicates (coefficients of variation of 26 and 31 percent, respectively). The disagreement between a batch equilibrium measurement of adsorption and one inferred from observing transport of an adsorbed chemical, a so-called dynamic method, is due in part to rate-limited transport of the organic chemical to isolated adsorption sites. Since these rate limitations are likely to be more extreme under natural conditions than when the structure is broken up, it seems essential to perform adsorption measurements in the field. However, to date no such adsorption measurements have been made, and information on chemical adsorption continues to be inferred from laboratory measurements and is usually based on the batch equilibrium method.

LIMITING MODEL ASSUMPTIONS

Because of the extreme limitations imposed by making measurements on the field scale under natural conditions, soil models of water and chemical transport are forced to assume a set of simplifying conditions which under certain circumstances may result in significant disagreement between prediction and observation. These model assumptions are summarized in Table 6-3 and will be reiterated below.

Virtually all models describing field-scale transport, except those with a natural multi-dimensional symmetry (e.g., water flow from a furrow) assume that water and chemical flow is one dimensional. This has the effect for large-scale simulations of requiring field averaged representative transport coefficients and a field-average retention curve. Although a one-dimensional field-scale model has had some success in predicting upward water flow and subsequent evapotranspiration [47], field-scale one-dimensional transport has not been demonstrated for downward flow for either water or chemical. In fact, Warrick et al. [48], illustrated that use of field-averaged transport coefficients in an average one-dimensional model would not predict the true mean behavior of a variable field. Instead, what must be done is to examine individual behavior at different parts of the field and average these results to estimate field-scale drainage. The main limitation to using a one-dimensional flow approximation is the size of the field and the extent of the variability of the parameters within the

Table 6-3. Limiting Transport Model Assumptions Imposed by Inadequate Measurements.

One-dimensional flow
Isothermal Soil
Non-Hysteretic Transport Functions
Homogeneous or Layer-Heterogeneous Soil
Equilibrium Non-Hysteretic Adsorption of Chemical
Complete Accessibility of Porous Medium (No immobile water)
Convective-Dispersive Transport of Chemical

field. Numerous current research efforts are under way to improve model representation of this problem (see review by Peck, [49]).

Virtually all water and chemical models assume isothermal soil conditions. The theory for coupling effects between water and temperature movement was worked out roughly by Philip and de Vries [50]. However, coupling effects require a host of transfer coefficients which are extremely difficult to measure and whose functional dependence on water content or temperature is largely unknown. For this reason, transport processes are usually considered isothermal, and a mean temperature value is used. Temperature effects are likely to be most significant in simulations of transport of volatile chemicals whose vapor pressure increases nonlinearly with temperature. Further, since the volatilization site is the soil surface where temperature variations are extreme, it is unclear the extent to which replacing such variations by an average value introduces error. A recent Ph.D. thesis [51] addresses this problem.

Non-hysteretic transport functions are assumed because of the extreme difficulty in obtaining a hysteretic representation of the relationship between two variables. Although the problem still generates intense academic interest, it seems unlikely in the future that hysteresis will be included in any transport model. Even with the aid of computer simulation, it is at the present time unclear the extent to which hysteresis is important in transport.

Because of the one-dimensional flow assumption, soil is necessarily assumed to be laterally homogeneous. This allows, at most, a heterogeneous layer representation for soil and does not permit lateral flow to occur. Since at progressively larger depths from the soil surface it becomes more and more difficult to calibrate transport coefficients, it seems likely in the future that representative transport coefficients will be assigned based on whatever geologic information is available about the strata being simulated.

Virtually all of the chemical transport models assume equilibrium non-hysteretic adsorption of the chemical. An equilibrium model could possibly be used in a rate-limited application provided that one obtained adsorption measurements under conditions of transport similar to those which will be simulated, i.e., at the same range of water fluxes. Such calibrations would automatically include the effect of rate-limited adsorption on the estimate of the adsorption coefficient. Adsorption hysteresis may be quite important for certain chemicals which desorb much less readily than they adsorb. This has the effect of enhancing their retention in layers where they are deposited. It seems likely that simple models of adsorption-desorption may be included in transport models in the future.

In most chemical transport models, it is assumed that the entire wetted pore space is available to the chemical, i.e., no immobile water. The major effect of making such an approximation is to represent the pore water velocity as the water flux divided by the total water content which could underestimate water velocity when substantial pore bypass exists. This assumption is certainly very inaccurate in regions where macropore effects are extreme; in such cases much of the water and chemical may be carried in a relatively small fraction of the pore space. There exist a number of mobile-immobile water models which include effects on chemical diffusion and adsorption that have been tested under laboratory conditions [52,53]. However, these models introduced a number of new parameters which are not directly accessible to measurement under field conditions. These parameters may be inferred only by applying the model to field data and by fitting the model to observation by minimizing the difference between the model and the data while varying parameters. Given the large number of factors which contribute to the disagreement between a model and an observation on the field scale, such minimization procedures are likely not to be useful in identifying subtle unmeasurable features of the transport process.

Analogous to the laboratory problem where most models have been developed and tested, virtually all chemical transport models used in the field assume convective-dispersive transport. Furthermore, most of these models are one dimensional, representing downward movement by a mean one-dimensional water flow and chemical mixing by an average dispersion coefficient. Field measurement of this dispersion coefficient is very difficult and introduces considerable error. Furthermore, there are fundamental objections to its use at least near the soil surface [31]. A recent set of solute transport models proposed for application near the soil surface (i.e., 0 to 3 m) do not employ a dispersion coefficient but use a purely convective model based on the presumption of substantial lateral variability in downward water flow [31,32]. In the interim, convective-dispersive models should probably be calibrated under conditions similar to those in which the simulation will be made in order to minimize the disagreement between the prediction and the observation.

RECENT IMPROVEMENT IN FIELD MEASUREMENT STRATEGY

Recently, the theory of regionalized variables [54] has been proposed as a method for maximizing the information content of a finite set of discrete soil samples taken on the field scale. This method, rather than using the classical statistical assumption of independence, investigates the spatial correlation between discrete measurements and constructs field autocorrelation functions to study the relationship of a measurement made at one place to another. The autocorrelation function is in turn used to generate a minimum variance estimate of the parameter at a place in which it was not measured using the information obtained from the measurements themselves. This method is called kriging and shows promise for improving the information content of a set of measurements. Illustrations of the kriging method are given in a recent article by Vieira et al. [55].

SUMMARY

The study of water and chemical transport on the field scale is a science in its infancy. Nevertheless, at this time there have been identified a significant number of soil, environmental, and management parameters which influence chemical transport through soil, and a large body of information has been assembled for finding the relationship between these parameters and the chemical transport. From years of study under controlled conditions, a substantial theoretical framework has been developed for describing water and chemical transport. This theory uses differential equations developed from a macroscopic mass balance and uses flux equations obtained as an extension to equations which are valid in one phase systems. Chemical transport generally involves liquid, vapor, and adsorbed phases so that a relationship between the phases is also used in addition to the transport equation. Although rate limitations can be expected to alter this equilibrium relation under general conditions, models frequently simplify this using linear equilibrium partition coefficients. For such systems, the three-phase transport may be replaced by an equivalent representation in terms of total concentration using equivalent three-phase transport velocities and dispersion coefficients.

Recent research in field-scale transport suggests that the extensive spatial variability of soil physical and chemical properties may require a new formulation for use when describing average transport across large crosssectional areas. Such research is only in the formulation stage right now and should be further verified before general conclusions can be drawn. In the interim, substantial field testing of existing and proposed models of transport should be conducted together with substantial calibration and validation. Only when a large body of experimental information is available will workers be able to further refine the models.

REFERENCES

1. Bear, J. Dynamics of Fluids in Porous Media (New York: American Elsevier, 1972).

2. Bigger, J. W., and D. R. Nielson. "Spatial Variability of the Leaching Characteristics of a Field Soil," Water Resour. Res. 12:78-84 (1976).

3. Nielsen, D. R., R. D. Jackson, J. W. Cary, and D. D. Evans. Soil Water (Madison, WI: American Society of Agronomy, 1972).

4. Millington, R. J., and J. M. Quirk. "Permeability of Porous Solids," Trans. Farady Soc. 57:1200-1207 (1961).

5. Jury, W. A., W. F. Spencer, and W. J. Farmer. "Use of Models for Predicting Relative Volatility Persistence and Mobility of Pesticides and Other Trace Organics in Soil Systems," in Hazard Assessment of Chemicals, Vol. 2, J. Saxena, Ed. (New York: Academic Press, 1983).

6. Spencer, W. F., and M. M. Cliath. "Desorption of Lindane from Soil as Related to Vapor Density," Soil Sci. Soc. Amer. Proc. 34:574-578 (1973).

7. Harper, L. A., A. W. White, R. R. Bruce, A. W. Thomas, and R. A. Leonard. "Soil and Microclimate Effects on Trifluralin Volatilization," J. Environ. Qual. 5:236-242 (1976).

8. Hillel, D. Soil and Water (New York: Academic Press, 1971).

9. Cosby, B. J., G. M. Hornberger, R. B. Clapp, and T. R. Ginn. "A Statistical Exploration of the Relationships of Soil Moisture Characteristics to the Physical Properties of Soils," Water Resour. Res. 6:682-690 (1984).

10. Green, R. E. "Pesticide-Clay-Water Interactions," in Pesticides in Soil and Water, W. D. Guenzi, Ed. (Madison, WI: Soil Science Society of America, 1974).

11. Hamaker, J. W., and J. M. Thompson. "Adsorption," in Organic Chemicals in the Soil Environment, C. A. I. Goring and J. W. Hamaker, Eds. (New York: Marcel-Dekker, Inc., 1972).

12. Gardner, W. R. "Solutions of the Flow Equation for the Drying of Soils and Other Porous Media," Soil Sci. Soc. Amer. Proc. 23:183-187 (1959).

13. Bird, R. B., W. E. Stewart, and E. N. Lightfoot. Transport Phenomena (New York: John Wiley & Sons, 1960).

14. Jury, W. A. "Simultaneous Transport of Heat and Moisture Through a Medium Sand," Ph.D. Thesis, University of Wisconsin (1973).

15. Biggar, J., and Cheung. "Adsorption of Picloram on Panoche, Ephrata and Palouse Soils," Soil Sci. Soc. Am. Proc 37:863-868 (1973).

16. Doorenbos, J., and W. O. Pruitt. "Crop Water Requirements," Irrigation and Drainage Paper 24 (Rome: FAO, 1976).

17. Shouse, P., W. A. Jury, L. H. Stolzy, and S. Dasberg. "Field Measurement and Modeling of Cowpea Water Use and Yield Under Stressed and Well Watered Conditions," Hilgardia 50:1-24 (1982).

18. Rao, P. S. C., and R. J. Wagenet. "Spatial Variability of Pesticides in Field Soils: Methods for Data Analysis and Consequences," Weed Sci. 33: (In press).

19. Taylor, H. H., and B. Klepper. "Water Uptake by Cotton Roots During an Irrigation Cycle," Aust. J. Biol. Sci. 24:853-859 (1971).

20. Richter, G. "A Microlysimeter and Field Study of Water and Chemical Movement Through Soil," MS Thesis, University of California, Riverside (1984).

21. El Abd, H. "Spatial Variability of the Pesticide Adsorption Coefficient," Ph.D. Thesis, University of California, Riverside (1984).

22. Rao, P. S. C., D. E. Rolston, R. E. Jessup, and J. M. Davidson. "Solute Transport in Aggregated Porous Media," Soil Sci. Soc. Amer. J. 44:1139-1146 (1980).

23. Wagenet, R. J., and J. J. Jurinak. "Spatial Variability of Solute Salt Content in Mankos Shale Watershed," Soil Sci. 126:342-349 (1978).

24. Russo, D., and E. Bresler. "Soil Hydraulic Processes as Stochastic Processes," Soil Sci. Soc. Amer. J. 45:682-687 (1980).

25. Gajem, Y. M., A. W. Warrick, and D. E. Myers. "Spatial Dependence of Physical Properties of a Typic Torrifluvent Soil," Soil Sci. Soc. Amer. J. 45:709-715 (1981).

26. Warrick, A. W., and D. R. Nielsen. "Spatial Variability of Soil Physical Properties in the Field," in Applications of Soil Physics (New York: Academic Press, 1980).

27. Beven, K., and P. Germann. "Macropores and Water Flow in Soils," Water Resour. Res. 18:1311-1325 (1982).

28. Jury, W. A. "Use of Solute Transport Models to Estimate Salt Balance Below Irrigated Cropland," Adv. Irrig. 1:87-104 (1982).

29. Cary, J. W. "Water Flux in Moist Soil: Thermal Versus Suction Gradients," Soil Sci. 100:168-175 (1965).

30. Jury, W. A., and E. E. Miller. "Measurement of the Transport Coefficients for Coupled Flow of Heat and Moisture in a Medium Sand," Soil Sci. Soc. Amer. Proc. 38:551-557 (1974).

31. Dagan G., and E. Bresler. "Solute Dispersion in Unsaturated Soil at Field Scale," Soil Sci. Soc. Amer. J. 43:461-466 (1979).

32. Jury, W. A., L. H. Stolzy, and P. Shouse. "A Field Test of the Transfer Function Model for Predicting Solute Transport," Water Resour. Res. 18:369-375 (1982).

33. Jury, W. A. "Field Scale Water and Solute Transport Through Unsaturated Soil," Proceedings of the International Conference on Soil Salinity Under Irrigation, Bet-Dagan, Israel (1984).

34. Day, P. R. "Particle Fractionation and Particle Size Analysis," in Methods of Soil Analysis, C. A. Black, Ed., ASA Monograph 9 (Madison, WI: American Society of Agronomy, 1965).

35. Taylor, S. A., and G. L. Ashcroft. Physical Edaphology (San Francisco, CA: Freeman and Co., 1977).

36. Allison, L. E. "Organic Carbon," in Methods of Soil Analysis, C. A. Black, Ed., ASA Monograph 9 (Madison, WI: American Society of Agronomy, 1965).

37. Blake, G. "Bulk Density" in Methods of Soil Analysis, C. A. Black, Ed., ASA Monograph 9 (Madison, WI: American Society of Agronomy, 1965).

38. Boersma, L. "Field Measurement of Hydraulic Conductivity," in Methods of Soil Analysis, C. A. Black, Ed., ASA Monograph 9 (Madison, WI: American Society of Agronomy, 1965).

39. Nielsen, D. R., J. W. Biggar, and K. T. Erh. "Spatial Variability of Field Measured Soil Water Properties," Hilgardia 42:215-259 (1973).

40. Rose, C. W., W. R. Stern, and J. E. Drummond. "Determination of Hydraulic Conductivity as a Function of Depth and Water Content for Soil in situ," Aust. J. Soil Res. 3:1-9 (1965).

41. Libardi, P. L., K. Reichardt, D. R. Nielsen, and J. W. Biggar. "Simple Field Methods for Estimating Soil Hydraulic Conductivity," Soil Sci. Soc. Amer. J. 44:3-6 (1980).

42. Sallam, A., W. A. Jury, and J. Letey. "Measurement of Gas Diffusion Coefficient Under Relatively Low Air-Filled Porosity," Soil Sci. Soc. Amer. J. 48:3-6 (1984).

43. Spencer, W. F., and M. M. Cliath. "Desorption of Lindane from Soil," Soil Sci. Soc. Amer. Proc. 34:574-578 (1970).

44. Spencer, W. F., and M. M. Cliath. Test Protocols for Environmental Fate and Movement of Toxicants (Arlington, VA: Association of Analytical Chemists, 1981).

45. Green, R. E., and J. C. Corey. "Pesticide Adsorption Measurement by Flow Equilibration and Subsequent Displacement," SSSA Proc. 35:561-565 (1971).

46. Jury, W. A., H. El Abd, and T. M. Collins. "Field Scale Transport of Nonadsorbing and Adsorbing Chemicals Applied to the Soil Surface," Proceedings of the NWWA U.S. EPA Conference on Characterization and Monitoring of the Vadose (Unsatruated) Zone, Las Vegas, NV (1983).

47. Nimah, M. N., and R. J. Hanks. "Model for Estimating Soil Water, Plant and Atmospheric Interrelations 1. Description and Sensitivity," Soil Sci. Soc. Amer. Proc. 37:522-527 (1973).

48. Warrick, A. W., G. J. Mullen, and D. R. Nielsen. "Scaling Field Measured Soil Hydraulic Properties Using a Similar Media Concept," Water Resour. Res. 13:355-362 (1977).

49. Peck, A. J. Advances in Irrigation, Vol. 2 (In Press, 1983).

50. Philip, J. R., and D. A. de Vries. "Moisture Movement in Porous Materials Under Temperature Gradients," Trans. Amer. Geophys. Union 38:222-232 (1957).

51. Streile, G. "Temperature Effects on Pesticide Transport," Ph.D. Thesis, University of California, Riverside (1984).

52. van Genuchten, M. Th., J. M. Davidson, and P. J. Wierenga. "An Evaluation of Kinetic and Equilibrium Equations for the Prediction of Pesticide Movement Through Porous Media," Soil Sci. Soc. Amer. Proc. 38:29-34 (1974).

53. van Genuchten, M. Th., and P. J. Wierenga. "Mass Transfer in Sorbing Porous Media," Soil Sci. Soc. Amer. J. 40:437-480 (1976).

54. Journel, A. G., and C. J. Huijbregts. Mining Geostatistics (London: Academic Press, 1978).

55. Vieira, S. R., J. R. H. Hatfield, D. R. Nielsen, and J. W. Biggar. "Geostatistical Theory and Application to Agronomy," Hilgardia 51:1-75 (1983).

CHAPTER 7

VOLATILIZATION FROM SOIL

W. A. Jury
University of California-Riverside, Riverside, California

PROCESS DESCRIPTION

Definition

Volatilization is defined as the loss of chemicals in vapor form from soil surfaces to the atmosphere. This process is ultimately limited by the chemical vapor concentration which is maintained at the soil surface and by the rate at which this vapor is carried away from the soil surface to the atmosphere. The potential volatility of a chemical is related to its inherent saturated vapor pressure, but the actual volatilization rate from soil in any specific circumstance will depend on all soil, atmospheric, or management factors which influence the behavior of the chemical at the soil-air-water interface [1].

Rate Process Interactions at the Soil Surface

It is useful to picture the volatilization process from soil as depending on the balance between two rate processes: the flux of chemical from the soil body to the soil surface and the flux of chemical vapor away from the soil surface to the atmosphere [2]. The relative sizes of these two rate processes determine whether or not chemical concentrations will build up or deplete in the surface layer and hence affect vapor density at the site of volatilization.

A number of soil, environmental, and management parameters influence the volatilization process through their influence on these two rate processes and the resulting soil vapor density at the surface. Prior to a detailed discussion of this influence, however, it is helpful to develop the fundamental descriptions of the soil chemical vapor density or concentration, the transport mechanisms to the soil surface, and the transport mechanisms from the soil surface to the air.

Soil Chemical Vapor Density

When a given quantity of organic chemical is mixed with moist soil, it ultimately partitions into adsorbed, solution, and vapor phases so that one may write the total quantity of chemical per unit soil volume (C_T) as

$$C_T = \rho_b C_A + \theta C_L + a C_G \tag{7-1}$$

where C_A is adsorbed chemical concentration (μg/g soil), C_L is dissolved chemical concentration (μg/cm^3 solution), C_G is vapor density (μg/cm^3/soil air), ρ_b is soil bulk density (g/cm^3), θ is volumetric water content, and a is volumetric air content. The units used in the individual phase concentrations are conventional so as to correspond with the usual methods of measurement.

The relationship between vapor density and associated solution concentration at equilibrium is usually given by Henry's Law

$$C_G = K_H C_L \tag{7-2}$$

where K_H is Henry's constant which in this system of units is dimensionless.

The applicability of Henry's law to soil systems has been confirmed for a variety of chemicals and circumstances [3,4,5,6]. Furthermore, Spencer and Cliath [6] showed that for the organic compounds they studied, the relationship equation (7-2) persists from very trace concentration levels all the way to saturation.

The equilibrium relationship between the solution concentration C_L and the adsorbed concentration C_A is called the adsorption isotherm

$$C_A = f(C_L) \tag{7-3}$$

where a variety of functional forms have been used to describe this relationship (see Chapter 8 on Adsorption). However, it is quite common, particularly for pesticides and other trace organics, to assume a linear relationship between adsorbed and liquid concentration called a distribution coefficient as defined in equation (7-4)

$$C_A = K_D C_L \tag{7-4}$$

where K_D is the distribution coefficient which in this system of units has a dimension of (cm^3/g). Combination of equations (7-1) through (7-4) allows a linear relationship to be formed between total organic chemical concentration and vapor concentration

$$C_T = [\rho_b K_D/K_H + \theta/K_H + a]C_G = R_G C_G \quad (7\text{-}5)$$

where R_G is the vapor partition coefficient as defined by Jury et al. [7].

The utility of the formulation given in equation (7-5) is that all soil, chemical, and environmental processes influencing the vapor concentration in equilibrium with a given total amount of chemical are contained in the quantities defined by the partition coefficent. To further separate the soil and chemical influences, the distribution coefficient K_D is sometimes expressed as

$$K_D = f_{oc}K_{oc} \quad (7\text{-}6)$$

where K_{oc} is the organic carbon partition coefficient, and f_{oc} is the organic carbon fraction. Trace organics and pesticides show a much smaller coefficient of variation in their K_{oc} coefficient between different soils than with their distribution coefficient K_D [8]. Thus, as an approximate index, one may take the organic carbon partition coefficient as a property of the chemical irrespective of the soil. The limitations of this formulation have been discussed in Jury et al. [7] and in Rao and Davidson [9].

Vapor Movement from Soil to Atmosphere

Although the motion of air in the atmosphere is generally turbulent, close to the soil surface there exists a relatively stagnant boundary layer through which gaseous vapor must move by molecular diffusion [10]. Jury et al. [7], represented this vapor flux away from the soil surface using Fick's law of diffusion through the boundary layer

$$J = \frac{D_G^{AIR}}{d}[C_G(o) - C_G(d)] \quad (7\text{-}7)$$

where D_G^{AIR} is the molecular diffusion coefficient of the gaseous chemical in air, $C_G(o)$ is the vapor concentration at the soil surface discussed in the previous section, and $C_G(d)$ is the gaseous concentration in the free air above the stagnant boundary layer. For most chemicals present in trace concentrations in soil, this free air concentration is small enough to be neglected compared to the surface concentration. The boundary layer thickness is an idealization to represent the degree of mixing of the surface layer. An alternative to equation (7-7) would be to use a transfer coefficient

$$h_d = D_G^{AIR}/d$$

between the surface and the atmosphere.

Chemical Movement from Soil to the Soil Surface

Chemicals present in the soil generally move by three mechanisms: (1) gaseous diffusion within air voids, (2) liquid diffusion within soil solution, and (3) mass flow of dissolved chemical within moving soil solution. These mechanisms were discussed in considerable detail in Chapter 6 and will be briefly summarized.

Gaseous diffusion. Movement of gases through soil is described by an extension of Fick's Law of diffusion

$$J_V = -\eta D_G^{AIR} \frac{\partial C_G}{\partial z} \tag{7-8}$$

where η is a tortuosity factor to account for reduced cross-sectional area and increased pathlength in soil. An empirical model devised by Millington and Quirk [11]

$$\eta = a^{10/3}/\phi^2 \tag{7-9}$$

where ϕ equals porosity, has received substantial experimental verification in soil over a large range of air contents, a [12,13,14, 15].

Liquid diffusion. In a manner similar to vapor diffusion in soil, a liquid diffusion equation is written as modification of Fick's Law

$$J_L = -\frac{\theta^{10/3}}{\phi^2} D_L^{WATER} \frac{\partial C_L}{\partial z} \tag{7-10}$$

where D_L^{WATER} is the diffusion coefficient of the chemical in pure water. The reduction factor used in equation (7-10) is again the Millington and Quirk model with the volumetric water content θ substituted for air content a.

Mass flow. The contribution of mass flow to chemical transport may be determined from the concentration of the chemical in solution and the water flux rate.

$$J_M = J_W C_L \tag{7-11}$$

where J_W is water flux rate. Thus, those chemicals with high solution concentration have potentially high mobility in soil where water is flowing.

Hydrodynamic dispersion. As discussed in Chapter 6, the water flux J_W is actually a volume-averaged quantity which neglects many of the complex three-dimensional water flow paths through the porous soil matrix. For this reason, equation (7-11) does not describe all of the solute mass flux, and a new term, the dispersion flux, J_{HD}, must be included in the description of transport. This term is given in equation (7-12).

$$J_{HD} = - D_{HD} \partial C_L/\partial z \qquad (7\text{-}12)$$

where D_{HD} is the hydrodynamic dispersion coefficient needed to describe the solute mixing due to mass flow at the pore scale.

The mathematical expressions above contain many parameters which are influenced by external factors in the soil environment. In the next section, these factors will be listed, and their influence on the volatilization process will be discussed in detail.

A major simplification in the general description of volatilization of soil-incorporated chemicals can be achieved by classifying them into two distinct groups: those whose resistance to volatilization is primarily in the soil (hereafter called category 1), and those whose resistance to volatilization is primarily in the stagnant boundary layer above the soil surface (hereafter called category 2). Jury et al. [7], showed that chemicals may be placed in category 1 if their Henry's constant K^H in the dimensionless units defined by equation (7-2) is greater than 10^{-5}. In qualitative terms, a chemical is considered to be in category 1 if material volatilizes to the soil atmosphere as rapidly as it is transported to the soil surface. Conversely, a chemical is in category 2 if the stagnant boundary layer acts as a partial barrier to transport which allows chemical concentrations to build up at the soil surface. Chemicals which belong in category 1 are sometimes called "well mixed."

Soil, Environmental, and Management Factors Influencing Chemical Volatilization from Soil

Table 7-1 summarizes the various parameters influencing chemical volatilization from soil divided, as in the case of the previous chapter, into groups of soil parameters, environmental parameters, and management parameters. This table is similar but not identical to Table 6-1 of the previous chapter. Several parameters discussed there are not included here because they have a negligible or undetermined influence on the volatilization process. In the following section, the effect of each of the parameters in Table 7-1 on the volatilization process will be discussed with specific emphasis on the critical transport mechanisms or partitioning processes involved in the rate process interactions at the soil surface. It will be helpful to keep in mind the volatilization classification developed in the previous section which entails grouping chemicals into the well-mixed category 1 if their Henry's constant $K_H > 10^{-5}$.

Table 7-1. Principal Soil, Environmental, and Management Parameters Influencing Chemical Volatilization from Soil.

Soil Parameters	Environmental Parameters	Environmental Parameters
Water content	Temperature	Chemical concentration
Bulk density or porosity	Wind	Depth of incorporation
Clay content	Evaporation	Irrigation management
adsorption site density	Precipitation	Soil management
Soil structure		

Soil Parameters

Soil water content. Volumetric soil water content (θ) influences the volatilization process primarily by affecting the gaseous and liquid diffusion coefficients which regulate transport of material upward to the soil surface. As indicated in equations (7-8) and (7-9) and (7-10), the soil diffusion coefficients depend non-linearly on volumetric water content or air content. Hence at higher water contents, transport by vapor diffusion drops dramatically while transport by liquid diffusion increases. However, many chemicals have non-negligible concentrations in both the liquid and vapor phases so that each transport mechanism can contribute to the movement of material upward to the volatilization surface. For this reason, Jury et al. [7], recommended looking at the effective vapor-liquid diffusion coefficient describing the total transport to the surface by diffusion in both phases. The mathematical development of this effective diffusion coefficient is described in Appendix A. Figure 7-1, taken from Jury et al. [7], shows the effective diffusion coefficient plotted as a function of water content for 20 organic pesticides which contain representatives of both category 1 and category 2 behavior. This figure reveals that many of the chemicals (in fact the category 2 chemicals) have a negligible vapor diffusion coefficient and thus have a steadily increasing effective diffusion coefficient with increasing water content (e.g., bromacil). For these chemicals, it is to be expected that transport by diffusion to the soil surface will increase dramatically with increasing water content. On the other hand, the category 1 chemicals with substantial vapor diffusion coefficients (e.g., lindane) will compensate for decreasing vapor transport by increasing liquid transport and hence will have less dependence on water content.

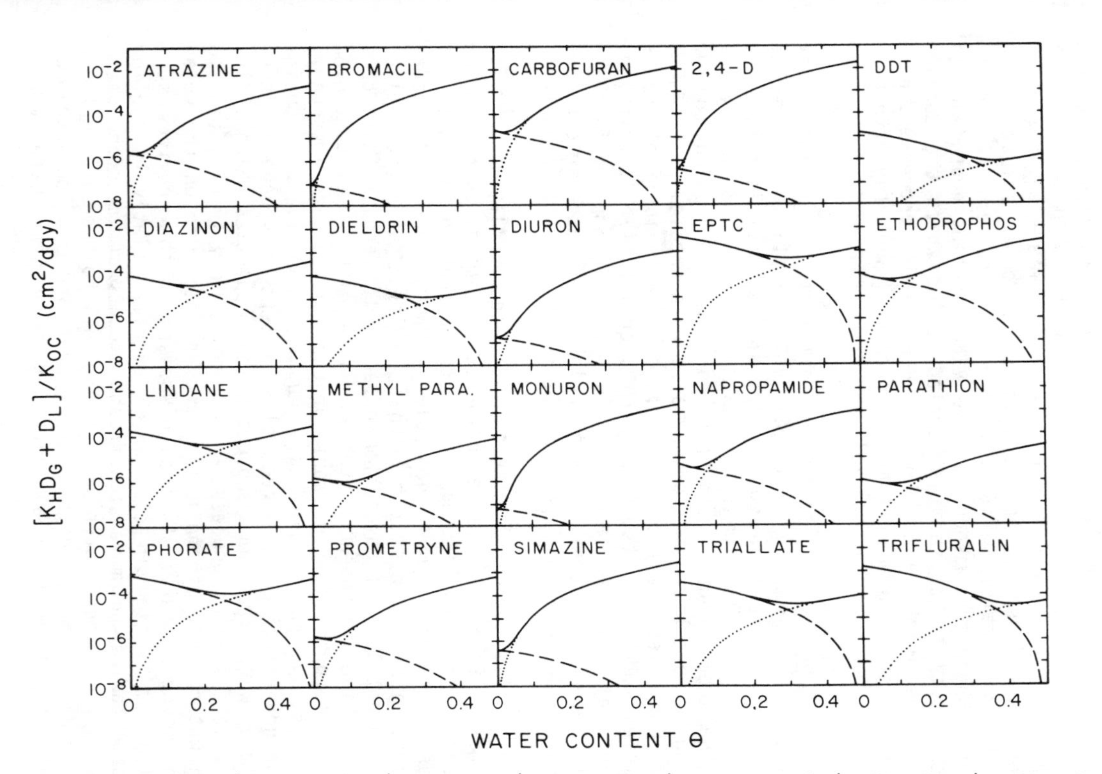

Figure 7-1. Calculated vapor (dashed line), liquid (dotted line), and total (solid line) effective diffusion coefficients as a function of water content for 20 chemicals. The effective diffusion coefficient combines vapor and liquid transport, reduced by adsorption (taken from Jury et al., [7].

A dramatic influence of water content on volatilization occurs when the soil is dried to a point where only a few monolayers of water remain on the soil solids. In this case, water molecules which preferentially occupy soil adsorption sites are removed, and the chemical adsorption capacity of the soil is greatly increased [16]. This increased adsorption capacity greatly decreases gaseous concentrations in the dry region, and volatilization drops to insignificant levels. However, when the soil surface layer rewets, the process is reversible, and much of the adsorbed solute is allowed to later volatilize [16,17]. For this reason, it may be possible to ignore this aspect of the water content dependence of adsorption provided that the time period during which the surface layer is dry is short-lived. For prolonged drying processes, however, when the soil surface is not allowed to rewet, the effect on volatilization could be substantial.

Bulk density or porosity. Decreasing soil porosity or increasing bulk density will generally decrease volatilization because the diffusive transport from the soil to the soil surface is reduced by decreasing the cross sectional area available for flow and by increasing the path length. Transport of chemical upward to the evaporating surface by mass flow may also be decreased if the soil hydraulic properties are reduced. This dependence is likely to be quite complex and lends itself only to a qualitative description. A secondary influence of decreasing soil porosity is to increase the density of soil mineral surfaces which could increase chemical adsorption and decrease transport to the surface by all mechanisms.

Clay content. The principal direct influence of clay content on chemical volatilization would be to increase chemical adsorption to mineral surfaces thereby decreasing the concentration in liquid and vapor phases. However, most non-polar organic chemicals adsorb primarily to soil organic matter so that no direct guidelines relating clay content to soil adsorption capacity for organic chemicals have been obtained in research studies. Also, ionic pesticides and trace organics, both of which would be expected to adsorb strongly to soil clays, generally have low vapor pressures and are considered to be non-volatile. However, clay content is strongly correlated to many water transport and retention properties which influence the volatilization process. Thus, soils high in clay tend to have high water contents and hence have an influence on the volatilization process similar to that discussed in the previous section. Also, soils high in clay which overlie relatively shallow ground-water tables can move significant amounts of water and chemicals upward from ground water or from the soil to the soil surface. Thus, the combination of high clay content and shallow depth to ground water could result in a significant transport upward by mass flow.

Adsorption site density. The adsorption site density refers to the quantity of adsorbing surfaces contributed from soil minerals or soil organic matter. The influence of adsorption on a soil-incorporated chemical is to increase the adsorbed and hence immobile phase concentration at the expense of the liquid and vapor phases.

Hence, increased adsorption always implies decreased volatilization because it decreases upward transport of chemicals from the soil to the surface to replace the materials lost by volatilization and because it also decreases the concentration of chemical vapor at the volatilization site. Jury et al. [7], demonstrated that the volatilization of category 2 chemicals, whose resistance to volatilization lies primarily in the boundary layer above the soil surface, is more strongly affected by increases in soil adsorption than is the volatilization of category 1 chemicals even though the latter are also affected by adsorption.

Environmental Parameters

Temperature. The influence of temperature on volatilization, although undoubtedly quite important, has received very little attention in the past. The discussion to follow, therefore, should be considered somewhat speculative as it is based on inference from the temperature dependence of the various processes influencing volatilization rather than on studies of the temperature dependence of volatilization itself.

For virtually all organic chemicals, saturated vapor density increases nonlinearly with increasing temperature [16]. However, the water solubility of certain chemicals will increase with increasing temperature whereas for others it will decrease [18]. Thus, the Henry's constant K_H given by equation (7-2) will strongly increase with increasing temperature thereby increasing the amount of vapor present for a given mass of chemical as temperature increases. Since the theoretical study of Jury et al. [7], predicted that volatilization will increase as Henry's constant increases, then it is to be expected that temperature increases will result in volatilization increases. As a general rule, chemicals will partition proportionally more of their total mass into the vapor phase at higher temperatures. Thus, chemical transport in the solution phase by mass flow and liquid diffusion will decrease compared to transport by vapor diffusion as temperature increases.

Wind. The influence of wind above the soil surface on volatilization is either to increase the mixing or equivalently to decrease the thickness of the stagnant boundary layer limiting volatilization. Since category 1 chemicals with large Henry's constants already act well mixed, wind speed will have only a moderate effect on volatilization for these chemicals. However, category 2 chemicals will have their volatilization rates increase dramatically as wind speed increases at the boundary layer. This phenomenon was demonstrated under field conditions by Glotfelty [19] who studied pesticide volatilization on two days with dramatically different wind speeds for the same chemicals.

Evaporation. The influence of water evaporation on organic chemical volatilization is quite complex as demonstrated by Jury et al. [7]. Figure 7-2 shows predicted volatilization fluxes for 20 chemicals during water evaporation rates of 0, 2.5, and 5 mm per

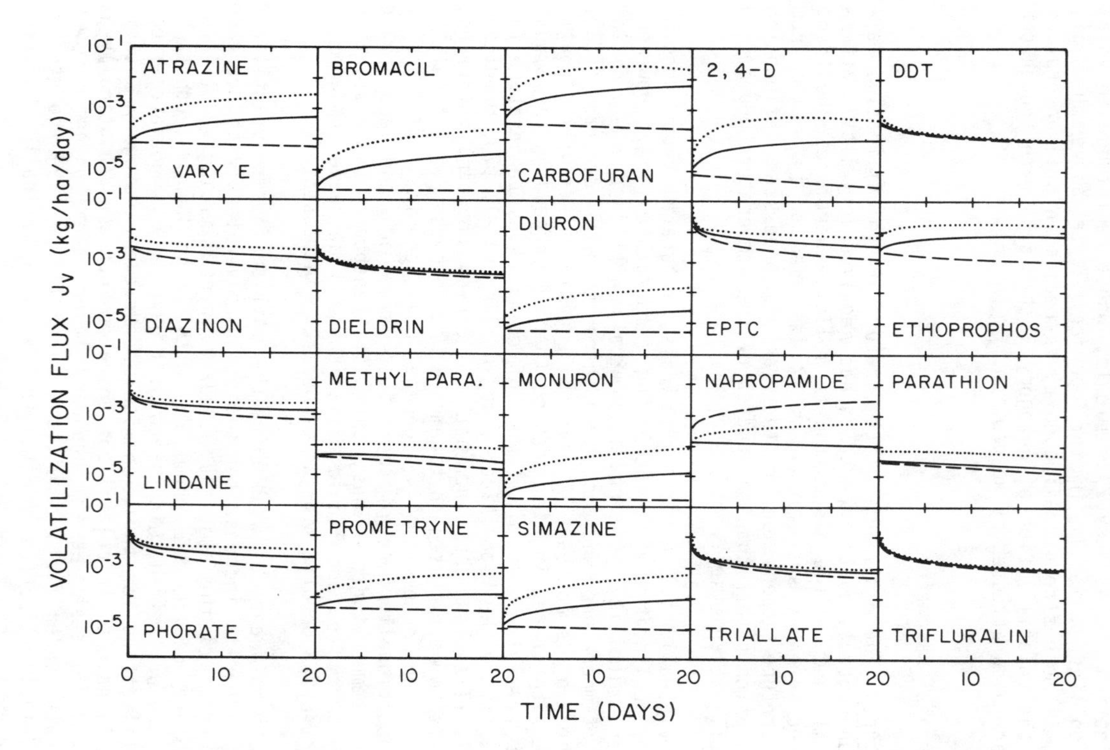

Figure 7-2. Volatilization fluxes predicted for 20 chemicals in soil at same concentration for three water evaporation rates E = 0 ---, E = 2.5 mm/day ____, E = 5.0 mm/day...(taken from Jury et al. [7]).

day. As a general rule, the category 2 chemicals such as atrazine, bromacil, etc. are more dramatically affected by water evaporation than are the category 1 chemicals such as DDT, dieldrin, etc., but other category 1 chemicals such as lindane and phorate will also have enhanced volatilization under water evaporation conditions. For all of the chemicals, soil water evaporation has the effect of carrying dissolved organic chemicals by mass flow to the soil surface where the volatilization process occurs. This enhances volatilization to varying degrees, depending on the extent of upward movement and the vapor pressure of the chemical at the volatilization site. The only chemicals which are completely unaffected by water evaporation are those such as DDT and trifluralin which are so strongly adsorbed that only insignificant amounts of upward transfer occur by mass flow.

Precipitation. The effect of precipitation on volatilization is indirect and serves to transport the chemical below the site of volatilization where soil layers above it greatly decrease its volatilization. Since it has no direct effect other than to transport the chemical away from the surface, rainfall is likely to be a more important limiter of volatilization for chemicals which leach readily and have low adsorption.

Management Parameters

Chemical concentration. Since the volatilization rate of a chemical is proportional to its vapor concentration at the soil surface, increasing chemical concentration creates higher vapor densities and hence higher volatilization. Indeed, if the chemical partitions linearly, then volatilization is proportional to chemical concentration.

Depth of incorporation. The volatilization rate of chemicals can be greatly decreased by incorporating them below the surface particularly if they are highly adsorbed [20]. Volatile chemicals which are sprayed on the soil surface are particularly prone to losses and should be plowed under soon after application. For a given mass of applied chemical, the depth of incorporation also has the effect of reducing the concentration and thus reducing volatilization as presented in the above argument.

Irrigation management. The effect of irrigation management is similar to rainfall in the sense of leaching chemicals below the volatilization site. In addition, the intensity and spacing between irrigations can affect the duration of the evaporation cycle which can have the effect of redepositing chemicals at the site of volatilization. Thus, high frequency irrigation would be likely to have the most significant effect on decreasing volatilization of incorporated chemicals.

Soil management. A significant amount of research has been conducted over the last century to study the effect of soil management on soil aeration for the purpose of maximizing diffusion of oxygen from the atmosphere down to the soil [21]. The same factors that increase diffusion of oxygen downward into the plant root zone will also increase the volatilization of chemicals from the soil to the atmosphere. Therefore, conditions of maximum aeration, i.e., soil with good structure, surface tilling, etc., will also tend to maximize the volatilization of chemicals.

PROCESS ROLE IN SOIL MODELS

Mayer et al. [22], discussed the influence of the upper boundary condition in volatilization predictions using five different models. Farmer et al. [14], developed and tested a model for predicting vapor loss of hexachlorobenzene from a landfill covered with soil. The model is based upon the vapor diffusion equations (7-8) and (7-9) applied in steady state through a soil cover. Jury et al. [13], developed and tested a volatilization model for triallate on two different soils. This model was able to describe time-dependent volatilization from a chamber initially filled to a uniform concentration with triallate and subsequently exposed to a moving air surface which maintained chemical concentration at a low value. The model combined the effects of liquid and vapor diffusion and liquid mass flow with a three-phase equilibrium model equation (7-1) using the convection-diffusion equation [23]. This model, when expressed in terms of total concentration, is given by equation (7-13)

$$\frac{\partial C_T}{\partial t} = D_E \frac{\partial^2 C_T}{\partial z^2} - V_E \frac{\partial C_T}{\partial z} \qquad (7\text{-}13)$$

This model has recently been proposed for use in screening large numbers of chemicals in Jury et al. [7]. Figure 7-2 shows predicted volatilization fluxes resulting from uniform initial concentrations for 20 different chemicals. Using the above theory, the influence of adsorption, water content, depth of incorporation, air flow rate, and other variables on volatilization rates for different chemicals are discussed in this document.

MEASUREMENT LIMITATIONS

All of the potential problems listed in Table 6-2 under the Implications heading apply equally well to volatilization measurements or modeling validation. Particularly imposing are the problems presented by macropores, cracks, etc. which are likely to create paths of low resistance for vapor transport to the atmosphere and which would not be well described by bulk tortuosity models such as those discussed above. Indeed, the influence of structural variability on vapor movement may be substantial because the gradient of vapor concentration generally points toward the soil surface so that the concentrated vapor in the soil would seek the path of least resistance to the atmosphere. The implications of this on field-scale transport are largely unexplored at this time because of the small number of field volatilization studies which have been conducted.

An additional problem to the ones discussed in the previous chapter is posed by the upper boundary condition for volatilization which assumes a stagnant boundary layer. This boundary layer is likely to be influenced by a variety of poorly characterized processes such as wind speed, surface roughness, and atmospheric turbulence above the zone of stagnation. The most severe limitation to use of this model for the upper boundary is the fact that the diffusion boundary layer thickness d in equation (7-7) is not directly measurable and must be inferred through an analogy. This will be discussed in the next section. Failure to include some boundary-layer limitation on volatilization, however, will likely result in a serious error when transport estimates are made on category 2 chemicals [24].

A potentially severe limitation to volatilization modeling under field conditions is the difficulty in estimating the effective adsorption of the chemical in a structured setting where many of the adsorption sites must be reached by a relatively slow diffusion process. In such cases, laboratory estimates of adsorption capacity are likely to poorly define the actual amount of adsorption occurring in the system [24].

METHODS OF FIELD COLLECTION OF PARAMETERS

Table 7-2 lists the parameters required to make volatilization estimates through modeling under field conditions divided into categories of soil, chemical, and atmospheric parameters. In some cases, information required to make these parameter estimates may be obtained under laboratory conditions. In other cases, the methodology required for field estimates is not available. These will be discussed below.

Table 7-2. Parameters Required for Volatilization Estimates in the Field.

Soil Parameters	Chemical Parameters	Atmospheric Parameters
Water flux	Henry's constant	Temperature
Liquid and vapor tortuosity functions	Diffusion coefficients in air and water	Boundary layer thickness
Water content	Adsorption parameters	
Bulk density	Degradation rate	

Parameters Affecting Volatilization

Soil Parameters

Water flux. Estimate of the water flux particularly for the case of upward flow toward evaporating surfaces may be obtained only for bulk average estimates valid for large areas. Meteorological methods such as those discussed by Tanner [25] may be used to estimate the potential evaporative loss when the soil surface is wet. When the soil surface dries, a simple rate-limited evaporation model may be used as discussed by Shouse et al. [26]. Models such as these will give estimates of average evaporation rate for time periods in between irrigations or rainfall and may serve as rough estimates of the potential for movement of dissolved chemical upward to the evaporating surface.

Liquid and vapor tortuosity functions. The tortuosity functions which model the reduction in diffusion of vapor or liquid due to the presence of the porous medium (e.g., equations 7-9, 7-10) have been measured almost exclusively under laboratory conditions. Because of the extreme complexity of three-dimensional chemical movement by diffusion, it is unlikely that reliable tortuosity function estimates can be made in the field. Instead, what is recommended is to use the Millington and Quirk [11] tortuosity functions discussed above (see equations 7-9, 7-10) which relate tortuosity to liquid or vaporfilled pore space.

Water content. Water content is required to estimate the tortuosity functions discussed above and is needed in the estimate of partitioning of chemicals into vapor and adsorbed phases. This parameter should probably be measured in the field setting by either neutron scattering methods or by gravimetric soil sampling.

Bulk density. Bulk density is needed to estimate the tortuosity function and needed to make estimates of the chemical partitioning into liquid, vapor, and adsorbed phases. It should be measured if possible in the field setting by taking undisturbed field cores whose density is measured in the laboratory. Porosity may be roughly calculated from bulk density.

Chemical Parameters

Henry's constant. The Henry's constant K_H may be set equal to the ratio of saturated vapor density to chemical solubility each of which may be measured by standard laboratory procedures [27]. There is no known method for directly measuring Henry's constant in the field, and at given temperature it should be considered a benchmark property of the chemical irrespective of the soil conditions. It is usually measured in the laboratory by setting it equal to the ratio of the saturated vapor density and the water solubility [6].

Diffusion coefficients in air and water. The pure chemical diffusion coefficients in air and in water have been measured for a restricted number of organic chemicals [7]. However, for intermediate weight molecular compounds, variation between chemicals is so slight that standard values of pure diffusion coefficients may be assumed to apply for all chemicals. Jury et al. [7], recommended the values 0.43 and 4300 cm^2/day for liquid diffusion coefficients in water and vapor diffusion coefficients in air, respectively.

Adsorption parameters. Considerable effort has been made in recent years to simplify the adsorption characterization of chemicals by using distribution coefficients K_D or even organic carbon coefficients K_{OC} to characterize the adsorption of a chemical. Information on these coefficients is generally available for chemicals either by direct measurement or through structural models or correlations with solubility. The more severe problem for field scale transport seems to be non-equilibrium adsorption due to preferential flow paths in the soil which bypass many of the adsorption sites which would be characterized by measurement. Indirect adsorption coefficients under field conditions could, in principle, be measured by comparing chemical leaching with that of a water tracer such as chloride, but this has not been done except under laboratory conditions [24]. In the interim, all that may be used to characterize the adsorption of a chemical is the standard adsorption coefficient or, in cases where an adsorption isotherm has been characterized, the Freundlich constants (see Chapter 8).

Degradation rate. Degradation rate affects chemical volatilization indirectly by decreasing the amount of chemical available for volatilization. In most cases, degradation rates under field conditions have been measured by mass balance estimates following intitial incorporation and therefore are likely to be contaminated by volatilization losses [7]. Summaries of effective degradation rate coefficients are available for a number of compounds [9,28], but these show large variation and merely serve as a benchmark estimate of persistence.

Atmospheric Parameters

Temperature. Soil temperature may be routinely measured by thermocouples or thermistors. Furthermore, if only mean temperatures are to be used, the average air temperature (max + min)/2 will give a reasonable estimate of soil temperature near the surface. The spatial variability of mean temperature is likely to be small and should not require extensive replication.

Boundary layer thickness. For chemicals which are not well mixed, the stagnant air boundary layer can act to limit volatilization. However, the boundary layer thickness which is required in several volatilization models [7,22] is not directly measurable. It may be inferred from atmospheric measurements of volatilization fluxes made by micrometeorological methods which measure the convection flux [19] provided that the soil surface concentration is known (see equation A-1 of Appendix A). Alternatively, as suggested by Jury et al. [7], it is assumed that the boundary layer for water vapor is the same as for other compounds. Then measurement of the evaporation rate above a wet surface and measurement of the surface temperature and air humidity define the boundary layer in terms of known parameters. This procedure leads to an estimate of a 0.5 cm boundary layer thickness for an evaporation rate of 0.5 cm/day at T = 25°C and a relative humidity of 50 percent. This value could be used for a rough estimate when no data are available.

REFERENCES

1. Spencer, W. F., W. J. Farmer, and W. A. Jury. "Review: Behavior of Organic Chemicals at Soil, Air, Water Interfaces as Related to Predicting the Transport and Volatilization of Organic Pollutants," Environ. Toxicol. Chem. 1:17-26 (1982).

2. Spencer, W. F., W. J. Farmer, and M. M. Cliath. "Pesticide Volatilization," Residue Rev. 49:1-47 (1973).

3. Call, F. "Soil Fumigation. V. Diffusion of EDB Through Soils," J. Sci. Food Agr. 8:143 (1957).

4. Goring, C. A. I. "Theory and Principles of Soil Fumigation," Adv. Pest Control Res. 5:47 (1962).

5. Leistra, M. "Distribution of 1,3-dichloropropene Over the Phases in Soil," J. Agr. Food Chem. 18:1124 (1970).

6. Spencer, W. F., and M. M. Cliath. "Desorption of Lindane from Soil," Soil Sci. Soc. Amer. Proc. 34:574-578 (1970).

7. Jury, W. A., W. F. Spencer, and W. J. Farmer. "Use of Models for Assessing Relative Volatility, Mobility, and Persistence of Pesticides and Other Trace Organics in Soil Systems," in Hazard Assessment of Chemicals, Vol. 2, J. Saxena, Ed. (New York: Academic Press, 1983).

8. Hamaker, J. W. and J. M. Thompson. "Adsorption," in Organic Chemicals in the Soil Environment, C. A. I. Goring, and J. W. Hamaker, Eds. (New York: Marcel-Dekker, Inc., 1972).

9. Rao, P. S. C., and J. M. Davidson. "Estimation of Pesticide Retention and Transformation Parameters." in: Environmental Impact of Nonpoint Source Pollution, M. R. Overcash and J. M. Davidson, Eds. (Ann Arbor, MI: Ann Arbor Science Publications, 1980).

10. Hartley, G. S. "Evaporation of Pesticides," Adv. Chem. Series 86:115-134 (1969).

11. Millington, R. J., and J. M. Quirk. "Permeability of Porous Solids," Trans. Faraday Soc. 57:1200-1207 (1961).

12. Letey, J., and W. J. Farmer. "Movement of Pesticides in Soil," in Pesticides in Soil and Water, W. D. Guenzi, Ed. (Madison, WI: Soil Science Society of America, 1974).

13. Jury, W. A., R. Grover, W. F. Spencer, and W. F. Farmer. "Modeling Vapor Losses of Soil-Incorporated Triallate," Soil Sci. Soc. Amer. J. 44:445-450 (1980).

14. Farmer, W. J., M. S. Yang, J. Letey, and W. F. Spencer. "Hexachlorobenzene: Its Vapor Pressure and Vapor Phase Diffusion in Soil," Soil Sci. Soc. Amer. J. 44:676-680 (1980).

15. Shearer, R. C., J. Letey, W. J. Farmer, and A. Klute. "Lindane Diffusion in Soil," Soil Sci. Soc. Amer. Proc. 37:189-194 (1973).

16. Spencer, W. F., and M. M. Cliath. "Pesticide Volatilization as Related to Water Loss from Soil," J. Environ. Qual. 2:284-289 (1973).

17. Harper, L. A., A. W. White, R. R. Bruce, A. W. Thomas, and R. A. Leonard. "Soil and Microclimate Effects on Trifuluralin Volatilization," J. Environ. Qual. 5:236-242 (1976).

18. Calvet, R. "Adsorption-desorption phenomena," in Interactions Between Herbicides and Soil, R. J. Hance Ed. (New York: Academic Press, 1980).

19. Glotfelty, D. D. "Atmospheric Dispersion of Pesticides from Treated Fields," Ph.D. Thesis, University of Maryland (1981).

20. Cliath, M. M., and W. F. Spencer. "Movement and Persistence of Dieldrin and Lindane in Soil as Influenced by Placement and Irrigation," Soil Sci. Soc. Amer. Proc 35:791-795 (1971).

21. Baver, L. D., W. H. Gardner, and W. R. Gardner. Soil Physics, 4th Ed. (New York: John Wiley & Sons, Inc., 1972).

22. Mayer, R. W., W. J. Farmer, and J. Letey. "Models for Predicting Volatilization of Soil Applied Pesticides," Soil Sci. Soc. Amer. Proc. 38:563-568 (1974).

23. Nielsen, D. R., R. D. Jackson, J. W. Cary, and D. D. Evans. "Soil Water," (Madison, WI: American Society of Agronomy, 1972).

24. Jury, W. A., H. El Abd, and T. M. Collins. "Field Scale Transport of Nonadsorbing and Adsorbing Chemicals Applied to the Soil Surface," Proceedings of the NWWA/U.S. EPA Conference on Characterization and Monitoring of the Vadose (Unsaturated) Zone, Las Vegas, NV (1983).

25. Tanner, C. B. "Evaporation of Water from Plants and Soil," in Water Deficits and Plant Growth, T. T. Kozlowski, Ed. (New York: Academic Press, 1968).

26. Shouse, P., W. A. Jury, L. H. Stolzy, and S. Dasberg. "Field Measurement and Modeling of Cowpea Water Use and Yield Under Stressed and Well-Watered Growth Conditions," Hilgardia 50: 1-25 (1982).

27. Freed, V. H. Literature Survey of Benchmark Pesticides. (Washington, DC: George Washington University Medical Center, 1976).

28. Nash, R. G. "Dissipation Rates of Pesticides from Soils," in CREAMS, Vol. 3, W. G. Knisel, Ed. (Washington, DC: U.S. Department of Agriculture, 1980).

CHAPTER 8

ADSORPTION OF ORGANIC CHEMICALS ONTO SOIL

W. A. Jury
University of California-Riverside, California

PROCESS DESCRIPTION

Definition

Adsorption refers to the bonding of a solute to adsorption sites on the soil solids, either soil mineral surfaces or organic matter surfaces. The effect of this bonding is to temporarily immobilize the molecule from transport in either the solution or vapor phase. In most quantitative descriptions of soil chemical transport processes in soil, the adsorbed molecules are represented as a separate phase, i.e., distinct from vapor or solution phases (see Chapter 6).

Practical Importance of Adsorption

The adsorption process is an extremely important aspect of chemical movement in soil because it acts to decrease chemical mobility in solution or vapor phases. For this reason, chemical adsorptive properties must be taken into account when designing fertilizer or pesticide application procedures, hazardous waste site management strategies, ground-water pollution potential estimates, etc. Since each chemical will, in principle, have a different adsorption affinity for solid surfaces in soil, it will move through soil in a unique manner in response to driving forces such as gravity. For this reason, quantitative description of the adsorption process is essential if descriptive models are to be developed for the transport of chemicals through the soil.

Published literature on adsorption of organic chemicals onto soil is quite voluminous and has been the subject of a number of excellent reviews written in recent years [1,2,3,4,5,6,7,8]. In spite of this widespread information, however, much of the knowledge of the adsorption mechanisms which bind organic chemicals in soil is empirical.

The uncertainty in the literature on adsorption to soils stems in large part from the diversity and complexity of the adsorption surfaces. Most organic chemicals are attracted to both clay mineral surfaces and organic matter surfaces. For a given chemical, the nature of the bonding mechanisms to the accessible adsorption sites depends on a variety of factors which will be discussed below.

DESCRIPTION OF BONDING MECHANISMS

Table 8-1 summarizes the principal molecular interactions involved in adsorption of organic chemicals to soil solid interfaces. Only a short description of each interaction will be given here. More details on the possible bonds between organic molecules and organic and mineral adsorbents can be found in Mortland [9], Calvet and Chassin [10], Theng [11], and Hayes [12].

Ionic Bonds

Adsorption by ionic bonds is due to ion exchange. Ionic bonds occur between organic anions or cations and positive or negative electric charges located at the adsorbent surfaces. Some organic molecules, such as certain herbicide molecules (e.g., paraquat or diquat), are always present as cations while others can be ionized with the extent of ionization depending on the acidity or pH of the medium. These latter compounds are either weak acids (e.g., phenoxyacetic acids) or weak bases (e.g., s-triazines).

Ligand Exchange

Ligand exchange interactions are possible in soil with the water molecule usually acting as the exchange ligand. Infrared spectroscopy investigations have verified this mechanism on montmorillonite clays [13]. Ligand exchange has also been postulated

Table 8-1. Intermolecular Interactions Involved in Adsorption.

High Energy Bonds	Low Energy Bonds
Ionic bonds	Charge-dipole and dipole-dipole bonds
Ligand bonds	Hydrogen bonds
	Charge transfer bonds
	van der Waal's-London bonds
	Entropy generation
	Magnetic bonds

for the binding onto the residual transition metals of humic acids [4].

Dipole Bonds

Charge-dipole bonds and dipole-dipole bonds are involved between polar organic molecules and electrically charged or polar adsorbing surfaces. This type of bonding has not been studied extensively for organic molecules in soil [2,7]. Thus, its relative importance in the adsorption process is unknown.

Hydrogen Bonds

Hydrogen bonds are mainly associated with NH or OH groups and N and O atoms. Thus, all organic molecules are able to establish hydrogen bonds. The hydrogen bond is important in the "water bridge" between an exchanger cation on a clay and a polar organic molecule [2]. This water bridge has been illustrated by Bailey et al. [14], and has been demonstrated to occur with clays for a number of different organic compounds containing nitrogen and/or oxygen [9]. At the present time, there is only limited evidence for the importance of hydrogen bonds in bonding of organic molecules in soil [7].

Charge Transfer Bonds

Charge transfer bonds are due to the transfer of electrons across the surface of the adsorbent or organic molecule. This mechanism has been postulated for organic cations adsorbed on clays [15] and for the adsorption of s-triazines onto organic matter [12].

van der Waal's Bonds

van der Waal's-London bonds are due to instantaneous dipoles established by fluctuations in the electron distributions as the electrons circulate in their orbitals. This interaction is weak and decreases as the sixth power of the inter-molecular distance. However, it can be a significant bond under certain circumstances. It has not been specifically studied in soil, but probably it is present in most organic chemical interactions with soil adsorption sites [7].

Entropy Generation

Entropy generation refers to the displacement of water molecules when organic molecules interact with the surface of an adsorbent, which increases the stability of the system [4]. It may be a contributing mechanism in the adsorption of nonpolar organic chemicals to hydrophobic surfaces such as soil humin.

Magnetic Interactions

Magnetic interactions occur from ring currents and conjugated double bonds and therefore may be a mechanism of adsorption for the larger organic molecules [4].

INFLUENCE OF SOIL, CHEMICAL, AND ENVIRONMENTAL PROPERTIES ON ADSORPTION

The complexity and diversity of the bonding mechanisms discussed in the previous section has obscured much of the correlation between adsorption and soil chemical or environmental properties. Table 8-2 summarizes the properties for which quantitative information is available. These will be discussed in greater detail below.

Properties Affecting Adsorption

Soil Properties

Clay composition and exchange capacity. Because clay mineral surfaces have a net negative charge, they will influence the adsorption reaction for virtually all chemicals. Their influence is strongest for compounds such as paraquat and diquat which carry a permanent positive charge, but they may also have a strong influence on chemicals which ionize in water. Thus, inorganic cations such as calcium, magnesium, and potassium, as well as positively charged trace metals such as cadmium, etc., are strongly adsorbed in soils with a large cation-exchange capacity. However, because of the large preponderance of organic matter surfaces and the non-polar nature of many organic chemicals, there is usually no correlation observed between adsorption and percent clay [2,7] unless the compound is permanently charged. A complete review of pesticide-clay-water interactions is given by Green [2].

Table 8-2. Soil, Chemical, and Environmental Properties Influencing Adsorption of Chemicals Onto Soil.

Soil Properties	Chemical Properties	Environmental Properties
Clay composition and exchange capacity	Electronic structure	Soil temperature
Organic matter content	Water solubility	
Soil water content	Solution composition	
Soil bulk density	Solution concentration	
	pH	

Organic matter content. Soil organic matter is an extremely complex and poorly defined substance. Because of this poor characterization, no exact relationships have been established between its structure and its adsorbing properties. However, it has been demonstrated that the adsorption of organic chemicals varies greatly according to the nature of organic matter, as demonstrated for trifuluralin by Grover [16] and for atrazine by Dunigan and MacIntosh [17]. Because of this dependence on organic matter type, it is likely that the adsorption capacity of a surface layer of soil will vary for a given total organic matter content depending on the state of decomposition of its residues [7].

All of these complications notwithstanding, there has been a positive linear relationship observed between soil organic matter content and adsorption of organic chemicals [4]. The correlation coefficient is likely to be highest when the organic matter content is large, but the correlation coefficient may still be significant even in soils with an organic carbon content as low as 0.1 percent [18]. Because of the positive correlation between soil organic matter and adsorption, workers have proposed defining an adsorption coefficient per unit of soil organic matter in order to reduce the variability between soils. This amounts to defining an organic carbon partition coefficient K_{OC} according to equation (8-1)

$$K_{OC} = K_D/f_{OC} \tag{8-1}$$

where f_{OC} is organic carbon fraction, and where K_D is the distribution coefficient discussed in Appendix A. As shown by Hamaker and Thompson [4], the coefficient of variability of the organic carbon partition coefficient particularly for nonpolar compounds is considerably less than the variation of the distribution coefficient for a given chemical adsorbing on different soils. However, as pointed out by Calvet [7], the remaining variability in the organic carbon partition coefficient is large enough to make it an unsuitable constant for precise analysis. However, it has still gained favor as a bench-mark property for soil organic chemicals [5,19,20] A good discussion of the limitations of partition coefficients may be found in Mingelgrin and Gerstl [8].

Soil water content. The soil water content can influence adsorption in two ways: it can modify the solution pathway leading to the adsorption sites and thus increase or decrease the accessibility of the surface to the solute [21], and it may also affect the physical-chemical properties of the adsorbent by increasing or decreasing the hydrolysis of clay lattices [22]. However, as shown by Spencer and Cliath [23] and Ehlers et al. [24], the influence of soil water content on adsorption is slight until the soil is extremely dry. When dry, the preferential coverage of water molecules on the soil's adsorbing surfaces is removed and solute adsorption increases dramatically. There is also some evidence that this effect is reversible when the soil rewets [23,25].

Bulk density. The influence of increasing soil bulk density on adsorption is to increase the density of adsorption sites per unit volume which will directly increase adsorption capacity. However, it is to be expected that the correlation between adsorption and bulk density for a group of soils will be small because clay and organic soils tend to be found at a lower bulk density than coarser-textured soils which are low in organic matter. Thus, the effect of increasing bulk density on adsorption refers to compressing a given soil volume.

Chemical Properties

Electronic structure. The electronic structure of the solute molecule has been found to be extremely important in adsorption. For example, if the molecule bears a permanent charge, such as paraquat or diquat, a strong binding mechanism of ion exchange to negatively charged mineral surfaces will be involved. On the other hand, neutral non-polar molecules such as ureas may bind to surfaces only by the van der Waal's interaction or perhaps through one of the other minor mechanisms discussed above. Although the bonding mechanisms discussed above and shown in Table 8-1 are quite complex, several integrated properties of the solute have been used to characterize the electronic structure and in turn have been correlated against adsorption. Among these are the Hammett sigma function and the Taft function [4] which describe the free energy changes accompanying bonding. Briggs [26] found a linear relationship between the organic carbon partition coefficient K_{OC} and the Hammett sigma function for 22 substituted phenyl-ureas.

Lambert [27] established a relationship between the parachor (related to the molecular volume) and the equilibrium adsorption coefficient for several phenyl-ureas and dialkylamines. This relationship was later modified by Hance [28] to include hydrogen bonding.

Water solubility. Since solute molecules compete with water for the adsorption sites, many workers have sought a relationship between adsorption and solubility. Kenaga and Goring [19] studied experimental data from 170 chemicals, mostly pesticides, and obtained regression coefficients between water solubility and organic carbon partition coefficients. These obtained correlation coefficients were all significant below the 1 percent level. Although, as pointed out by Calvet [7], a general relationship between water solubility and adsorption has not been established for all chemicals, the correlation appears to be strong enough within a class of compounds to provide useful regression equations between solubility and adsorption.

Solution composition and concentration. The complexity of the soil solution is such that very few quantitative relationships have been established between adsorption and macroscopic properties characterizing the soil solution in terms of ionic strength, electrical conductivity, pH, etc. For inorganic cations such as calcium, magnesium, and potassium, quantitative relationships have been established between cation activity which is a function of soil or ionic strength and exchange adsorption. It is to be expected that other positively charged compounds will also experience competitive adsorption effects which depend on the composition and concentration of the soil solution. However, when nonionic chemicals are considered, the influence of bulk soil solution parameters is generally neglected and the distribution coefficient or organic carbon partition coefficient is considered to be independent of these effects. Organic solvents can also have a large influence on adsorption and are frequently used to apply chemicals in a highly concentrated state which can greatly exceed their solubility in pure water.

pH. The pH of the soil system can have a marked effect on adsorption of compounds in soil, particularly compounds which are weakly acidic or weakly basic. Weak acids are in the free acid form when the pH of the soil solution is low, and they are much more highly adsorbed in this form than when present as an anion. Weak bases are converted to cationic forms at low pH which are also more highly adsorbed than as free bases. Polar materials may not actually form cationic or anionic forms but are capable of forming hydrogen bonds which may depend somewhat on pH. On the other hand, highly neutral molecules such as chlorinated hydrocarbons would be expected to show virtually no dependence on pH [4].

Despite these dependencies, however, the correlation of pH with adsorption in soil has proved to be much lower than the correlation between organic matter and adsorption or between cation exchange capacity and adsorption [29]. As a general rule, chemicals which dissociate in solution are likely to be strongly affected by pH whereas the rest will appear to be relatively independent of pH [1,2].

Environmental Properties

Soil temperature. Although some compounds may increase their adsorption as temperature increases [7], the majority decrease [18]. Temperature variations have two distinct effects on the adsorption process: solute-surface interaction effects and water-solute interaction effects. The balance between these two effects will determine the direction of dependence of the temperature. As a general rule, the effect of temperature on adsorption equilibrium is a direct indication of the strength of adsorption so that the weaker the bond the less the influence of temperature [4]. Summaries for the temperature effect of adsorption for a number of chemicals are given in the review article by Calvet [7].

PRACTICAL LIMITATIONS TO APPLYING ADSORPTION MODELS TO FIELD STUDIES

Because of the extreme complexity of the soil solution and the heterogeneous nature of the adsorption sites in soil, it seems unlikely that general relationsips will be developed in the immediate future describing in detail the dependence of adsorption on the variety of physical and chemical properties discussed above. For this reason, engineering approximations such as the distribution coefficient or the partition coefficient have gained acceptance as approximate methods for describing the adsorption capacity of a given chemical. There are certain complications which should be borne in mind for using these coefficients in chemical modeling, the most severe of which are discussed below.

Adsorption-Desorption Hysteresis

Virtually all organic chemicals studied and many inorganic compounds show a non-singular adsorption-desorption isotherm and exhibit greater resistance to desorbing than to sorbing. van Genuchten et al. [30], attempted to take this into account in modeling movement of pulses of chemical through soil columns. The major effect of this increased desorption resistance is to leave higher residual concentrations of chemicals in soil during leaching than would be predicted from the adsorption properties alone. However, the extent of this effect is still unknown. Resolution of the influence of hysteresis on transport is difficult because its effect on the shape of a column breakthrough curve is similar to that caused by rate-limited adsorption-desorption without hysteresis. Some different adsorption-desorption models are discussed in Rao and Davidson [5].

Influence of Soil Structure

The major method for characterizing soil adsorption is a batch equilibrium determination of the adsorption isotherm. In this method, all of the organic and mineral surfaces are exposed to the solute by shaking, and the change in solution concentration is measured until equilibrium is reached. Under natural conditions, however, soil is likely to be aggregated or cracked with certain of the surfaces inaccessible to the solute except by the time-consuming diffusive transport. For this reason, the amount of adsorption occurring under natural soil-aggregated conditions with flowing solution is likely to be less than that predicted from batch determination, and the adsorption process itself is likely to be rate-limited [31]. Since the rate coefficient used to describe the approach to equilibrium under rate-limited conditions is not at the present time predictable from bulk soil properties, no *a priori* modeling predictions can be made. However, recent attempts have been made to describe rate-limited diffusive transport by using a spherical aggregate structural model [32].

Influence of Spatial Variability

On the field scale, i.e., agricultural fields or large waste disposal areas, order of magnitude lateral variation has been observed in the primary soil properties influencing adsorption

[31,33,34]. For this reason, it is expected that the distribution coefficient K_D will have a value which varies from point to point in a large field either due to variations in soil organic matter or to a complex interaction between soil structure and transport. It thus should be concluded that prediction of the field wide average and extreme behavior of a chemical moving through natural soil is at this time an unresolved research problem [35]. However, at any particular location in the field, measurement of the distribution coefficient may be effective in modeling movement of the chemical in the immediate vicinity of the measurement.

Nonequilibrium Adsorption

When chemicals are flowing through soil, the residence time of an adsorbing solute may be too short to reach equilibrium with the adsorbing surface. Under such conditions, adsorption is less than predicted from equilibrium parameters such as K_D or K_{OC}. Rate-limited models (see Appendix A) have been proposed to account for this effect, but they contain parameters which cannot be independently measured.

Linear Adsorption Models

As mentioned above, the complications of adsorption are frequently neglected in favor of using the simple distribution or partition coefficient K_D describing the relationship between adsorbed and solute concentrations. Because of the enormously large number of organic chemicals on the market or under development, this simple index of adsorption has gained favor as a bench-mark property for characterizing potential mobility of chemicals under leaching [36]. Large compendiums of K_D or K_{OC} values have been published in the literature [4,5,37] including in certain cases the coefficient of variation found for the adsorption coefficient among different soils. As mentioned above, the K_{OC} coefficient has been observed to have less variability among different soils than the K_D in most cases.

As a further simplification in representing adsorption, it has been proposed to express the adsorption of all chemicals in terms of an octanol-water partition coefficient K_{OW}. Highly significant correlations have been found between K_{OW} and K_{OC} [5,38]. However, because this relationship is between the logarithms of the coefficients, prediction of adsorption potential from a K_{OW} measurement may involve order of magnitude error [19].

As shown in the chapter on chemical transport, use of a linear partitioning relationship between solution and adsorbed concentrations allows considerable simplification in mathematical modeling. For this reason, adsorption isotherms are frequently linearized over the range expected to be found in a given simulation. As demonstrated by Hamaker and Thompson [4], no single K_D value can describe adsorbed-solute partitioning from trace concentrations all the way to very high concentrations. Hence, if a single K_D value is used at both low and high concentrations, models predicting transport will give erroneous descriptions [5]. However, Karickhoff et al. [39], showed that the linear representation was a good approximation to the isotherm for trace concentrations of a chemical which are frequently at the levels found in soil.

Briggs [26] and others have developed a considerable amount of data relating adsorption to chemical structure which involves correlations between water solubility and adsorption and which involves molecular properties such as melting point, etc. Although these regressions have been found to be useful for a limited number of chemicals, they do not have a general validity and should be used with caution.

FIELD MEASUREMENT OF ADSORPTION PARAMETERS

Virtually all adsorption studies have been conducted under laboratory conditions. The only information obtained to date on field adsorption studies consists of observing transport of an adsorbed chemical. As shown by El Abd [31], simultaneous observation of the movement of a mobile tracer such as chloride or bromide and an adsorbed compound can be used to calculate an effective adsorption coefficient for the compound. For this simple procedure, all that is required is a measurement of the chemical concentration which may be obtained from soil core measurements. In principal, soil core measurements may also be made of chemical concentrations for the purpose of calculating adsorption by the method of El Abd [31]. However, adsorbed compounds are frequently difficult to extract through solution samples particularly if strongly adsorbed. Therefore, the soil coring method is frequently preferable. The more complex adsorption functional relationships discussed in Appendix A have never been characterized under field conditions. These methods require precise laboratory procedures for preparation and extraction in order to give the quantitative information required to determine the functional parameters. In the future, it is likely that simple partitioning relationships will be used under field conditions.

REFERENCES

1. Bailey, G. W., and J. L. White. "Factors Influencing the Adsorption, Desorption and Movement of Pesticides in Soil," Residue Rev. 32:29-92 (1970).

2. Green, R. E. "Pesticide-Clay-Water Interactions," in Pesticides in Soil and Water, W. D. Guenzi, Ed. (Madison, WI: Soil Science Society of America 1974).

3. Weed, S. B., and B. Webber. "Pesticide-Organic Matter Interactions," in Pesticides in Soil and Water, W. D. Guenzi, Ed. (Madison, WI: Soil Science Society of America, 1974).

4. Hamaker, J. W., and J. M. Thompson. "Adsorption," in Organic Chemicals in the Soil Environment, C. A. I. Goring and J. W. Hamaker, Eds. (New York: Marcel-Dekker, Inc., 1972).

5. Rao, P. S. C., and J. M. Davidson. "Estimation of Pesticide Retention and Transformation Parameters," in Environmental Impact of Nonpoint Source Pollution, M. R. Overcash and J. M. Davidson, Eds. (Ann Arbor, MI: Ann Arbor Science Publications, 1980).

6. Karickhoff, S. W. "Semi-Empirical Estimation of Sorption of Hydrophobic Pollutants on Natural Sediments," Chemisphere 10:833-846 (1981).

7. Calvet, R. "Adsorption-Desorption Phenomena," in Interactions Between Herbicides and Soil, R. J. Hance, Ed. (New York: Academic Press, 1980).

8. Mingelgrin, U., and Z. Gerstl. "Reevaluation of Partitioning as a Mechanism of Nonionic Chemical Adsorption in Soil," J. Environ. Qual. 12:1-11 (1983).

9. Mortland, M. M. "Clay Organic Complexes and Interactions," Adv. Agron. 22:75-114 (1970).

10. Calvet, R., and P. Chassin. "Complexes organiques des argile Mechanismes de formation," Bul. Groupe Fr. Argile 21:87-113 (1973).

11. Theng, B. K. G. The Chemistry of Clay-Organic Reactions (London: Hilger, 1974).

12. Hayes, M. H. B. "Adsorption of Triazine Herbicides on Soil Organic Matter," Residue Rev. 32:131-168 (1970).

13. Mortland, M. M., and W. F. Meggitt. "Interaction of EPTC with Montmorillonite," J. Agr. Food Chem. 14:126-129 (1966).

14. Bailey, G. W., and J. L. White, and T. Roghberg. "Adsorption of Organic Herbicides by Montmorillonite: Role of pH and Chemical Character of the Adsorbate," Soil Sci. Soc. Amer. Proc. 32:222-234 (1968).

15. Haque, R., S. Lilley, and W. R. Coshow. "Mechanism of Adsorption of Diquat and Paraquat on Montmorillonite Surfaces," J. Colloid Int. Sci. 3:185-188 (1970).

16. Grover, R. "Adsorption and Desorption of Trifluralin, Triallate, and Diallate by Various Adsorbents," Weed Sci. 22:405-408 (1974).

17. Dunigan, E. P. and T. H. MacIntosh. "Atrazine-Soil Organic Matter Interaction," Weed Sci. 19:279-282 (1971).

18. Lyman, W. J. "Adsorption Coefficients in Soil and Sediments," in Handbook of Chemical Property Estimations Methods, W. J. Lyman, Ed. (New York: McGraw-Hill, 1982).

19. Kenaga, E. E., and C. A. I. Goring. "Relationship Between Water Solubility, Soil Sorption, Octanol-Water Partitioning, and Bioconcentration of Chemicals in Biota," Proceedings of the 3rd ASTM Symposium on Aquatic Toxicology, ASTM Special Technical Publication 707:78-115 (1980).

20. Jury, W. A., W. F. Spencer and W. J. Farmer. "Use of Models for Assessing Relative Volatility, Mobility and Persistence of Pesticides and Other trace Organics in Soil Systems," in Hazard Assessment of Chemicals, Vol. 2, J. Saxena, Ed. (New York: Academic Press, 1983).

21. Grover, R., and R. J. Hance. "Effect of Ratio of Soil to Water on Adsorption of Linuron and Atrazine," Soil Sci. 109: 136-138 (1970).

22. Frenkel, H., and D. Suarez. "Hydrolysis and Decomposition of Calcium Montmorillonite," Soil Sci. Soc. Amer. Proc. 41:887-891 (1977).

23. Spencer, W. F., and M. M. Cliath. "Desorption of Lindane from Soil as Related to Vapor Density," Soil Sci. Soc. Amer. Proc. 34:574-578 (1973).

24. Ehlers, W., W. J. Farmer, W. F. Spencer, and J. Letey. "Lindane Diffusion in Soils. II. Water Content, Bulk Density and Temperature Effects," Soil Sci. Soc. Amer. Proc. 33:504-509 (1973).

25. Harper, L. A., A. W. White, R. R. Bruce, A. W. Thomas, and R. A. Leonard. "Soil and Microclimate Effects on Trifuluralin Volatilization," J. Environ. Qual. 5:236-242 (1976).

26. Briggs, G. G. "Molecular Structure of Herbicides and their Sorption by Soils," Nature 223:1288-1289 (1969).

27. Lambert, S. M. "Functional Relationships between Sorption in Soil and Chemical Structures," J. Agr. Food Chem. 15:572-576 (1967).

28. Hance, R. J. "An Empirical Relation Between Chemical Structure and the Sorption of Some Herbicides by Soil," J. Agr. Food Chem. 17:630-631 (1969).

29. Farmer, W. J., and Y. Aochi. "Picloram Sorption by Soils," Soil Sci. Soc. Amer. Proc. 38:418-423 (1972).

30. van Genuchten, M. Th., J. M. Davidson, and P. J. Wierenga. "An Evaluation of Kinetic and Equilibrium Equations for the Prediction of Pesticide Movement Through Porous Media," Soil Sci. Soc. Amer. Proc. 38:29-34 (1974).

31. El Abd, H. "The Spatial Variability of the Pesticide Adsorption Coefficient," Ph.D. Thesis, University of California, Riverside (1984).

32. Rao, P. S. C., J. M. Davidson, R. E. Jessup, and H. M. Selim. "Evaluation of Conceptual Models for Describing Nonequilibrium Adsorption-Desorption of Pesticides," Soil Sci. Soc. Amer. J. 43:22-28 (1979).

33. Gajem, Y. M., A. W. Warrick, and D. E. Myers. "Spatial Dependence of Physical Properties of Typic Torrifluven Soil," Soil Sci. Soc. Amer. J. 45:709-715 (1981).

34. Nielsen, D. R., J. W. Biggar, and K. T. Erh. "Spatial Variability of Field Measured Soil Water Properties," Hilgardia 42:215-259 (1973).

35. Jury, W. A. "Field Scale Water and Solute Transport Through Unsaturated Soil," Proceedings of the International Conference on Soil Salinity Under Irrigation, Bet-Dagan, Israel (1984).

36. McCall, P. J., R. L. Swann, E. A. Laskowski, S. M. Unger, S. A. Vrona, and H. J. Dishburger. "Estimation of Chemical Mobility in Soil from HPLC Retention Times," Bull. Env. Contam. Toxicol. 24:190-195 (1980).

37. Kenaga, E. E. "Predicted Bioconcentration Factors and Soil Sorption Coefficients of Pesticides and Other Chemicals," Etox. and Environ. Safety 4:26-38 (1980).

38. Karickhoff, S. W., and D. S. Brown. "Paraquat Sorption as a Function of Particle Size in Natural Sediments," J. Environ. Qual. 7:246-252 (1978).

39. Karickhoff, S. W., D. S. Brown, and T. A. Scott. "Sorption of Hydrophobic Pollutants on Natural Sediments and Soils," Water Res. 13:241-248 (1979).

CHAPTER 9

BIOTRANSFORMATION

R. L. Valentine and J. L. Schnoor
University of Iowa, Iowa City, Iowa

PROCESS DESCRIPTION

Definition

The term biotransformation is a general term describing any alteration of a compound affected by living organisms. Biodegradation is a more specific term usually referring to a biologically mediated transformation of a chemical into more simple products by the removal of one or more substituent groups. Neither of the above terms expresses the extent of change in the identity of a compound, the mechanisms involved in that change, the rate at which this change occurs, or the species responsible for this change.

Other terminology is frequently used to describe biodegradation. Mineralization refers to the degradation of a chemical to inorganic products such as carbon dioxide, water, ammonia, sulfate, nitrate, and chloride. The term ultimate degradation is frequently used interchangeably with mineralization. Partial degradation is commonly used to describe a level of degradation less than complete mineralization. Compounds that are not easily degraded are recalcitrant and persistent in the environment [1,2].

Importance of Microorganisms

In the soil environment, biotransformation of a compound is a result of chemical reactions catalyzed by enzymes which are produced as part of the metabolic activity of living organisms. Biotransformation in the soil has been primarily attributed to the action of microorganisms although exudates from plants and excretia from higher life forms may contribute to the overall degradation of a compound.

It is generally believed that bacteria, actinomycetes, and fungi are of primary importance. Algae have also been implicated in the degradation of some chemicals [3,4] although their importance is probably small except near the soil surface. The species of organisms in the soil are similar to those of an aquatic environment [5]. The collective sum of all microorganisms present is sometimes referred to as a consortia which constitutes a mixed population.

The bacteria are usually the most abundant group on a numerical basis. However, because bacteria are small and actinomycetes and fungi are large, frequently having extensive filaments, bacteria may not account for the majority of microbiological mass in a soil [2]. The actinomycetes which are classified as filamentous bacteria are a transitional group between bacteria and fungi and are usually second to bacteria in abundance [2]. However the relative importance of each group and species within each group depends on many factors such as oxygen and hydrogen ion concentration (pH).

Nutritional and Metabolic Considerations

Microorganisms require an energy and carbon source in addition to various nutrients to maintain themselves. Microorganisms are generally placed in one of two broad categories. Organic compounds are biodegraded principally by heterotrophic organisms which require organic compounds to serve as both a source of energy and carbon for cell growth and biosynthetic reactions. Fungi, actinomycetes, protozoans, and most bacteria are heterotrophic. Autotrophic organisms obtain energy from the oxidation of inorganic compounds such as ammonia, or from sunlight, and use carbon dioxide as a sole carbon source. Algae are primarily autotrophic although several species can also oxidize organic carbon to replace light [2].

The type of terminal electron acceptor (hydrogen acceptor) used in the electron transport system which extracts energy via oxidation of substances is an important characteristic of microbial metabolism. Under aerobic conditions, oxygen is used. Organisms growing in the absence of oxygen must use an organic product of metabolism (metabolic intermediate) or some inorganic substance such as nitrate or sulfate as an electron acceptor.

Microorganisms can also be classified based on their ability to grow in the presence or absence of oxygen. Three distinct groups exist [2]: aerobes require oxygen and do not grow in its absence; anaerobes can grow only in the absence of oxygen; and facultative anaerobes can grow either in the absence or presence of oxygen by switching electron acceptors. The ultimate product of aerobic respiration is carbon dioxide and water. Anaerobic metabolism produces incompletely oxidized simple organic substances such as organic acids in addition to several other products such as carbon dioxide, water, methane, and hydrogen gas.

Microbial growth frequently requires several other minerals and organic substances. Nitrogen, phosphorus, potassium, sulfur, magnesium, zinc, calcium, manganese, copper, cobalt, iron, and molybdenum are typical elements found in microorganisms. Some of these minerals such as molybdenum are required in extremely small amounts. Microorganisms may also need externally supplied growth factors which are organic molecules not synthesized by the organism but required in trace quantities for growth. Growth factors are structural building blocks such as amino acids and vitamins.

The biotransformation of a compound is frequently a stepwise process involving many enzymes and many species of organisms. Enzymes show a marked specificity in catalyzing a single type of reaction of one or a few related substrates. Therefore, extensive degradation of a complex substrate requires a large number of enzymes. A single species of organism may degrade a compound by the sequential breakdown of intermediate products. Several intermediates may be produced, and some of these intermediate products may not be further degraded. Consortia of bacteria may operate in a manner in which metabolites produced by one species are used as substrates for another.

Biotransformations may occur either inside a microorganism via intracellular enzymes or outside the organism by the action of extracellular enzymes. Soil enzymes may also be released as a result of cell lysis, and these enzymes may retain their activity in the soil. Although the role of extracellular enzymes in the transformation of xenobiotic compounds is unclear, such reactions are known to occur [6]. Getzin and Rosefield [7,8] have shown that an extracellular enzyme found in soil was capable of degrading malathion.

Enzymes always present in a microorganism and part of the "normal" metabolic activity are called constitutive enzymes. Other enzymes may be produced in response to a specific substrate and are called inducible enzymes. If an inducible enzyme is required for the degradation of a substrate, then significant degradation will occur only after a sufficient time period (time lag or acclimation period) has elapsed during which the inducible enzyme has been produced to an adequate concentration.

Frequently, the biodegradation of a substrate does not lead to energy production or formation of an essential nutrient. In other words, the substrate does not promote growth. Such substrates cannot be used as sole carbon and energy sources. Biodegradation of this sort has been termed cometabolism [9] or co-oxidation if the transformation involves an oxidation [10] and probably occurs widely in the environment [9,11]. Cometabolism occurs when an enzyme produced by an organism to degrade one substance also degrades another although the metabolic intermediates cannot be used. The metabolic intermediates are usually similar to the substance cometabolized and accumulate in the environment.

At low substrate concentrations, the organism may not get enough energy to supply maintenance requirements even if the compound can be used as an energy source. This could result in a persistent compound. However, low concentrations of a substrate may be degraded when another source of energy and carbon is available. The phenomena is called secondary utilization since the trace material does not significantly contribute to growth and maintenance of the organism [12]. Secondary utilization of substrates may be important in the degradation of trace concentrations of chemicals.

BIOLOGICAL REACTIONS

The pathways for the biological transformation of many anthropogenic organic chemicals have been determined. These chemicals include hydrocarbons, pesticides, herbicides, and fungicides. Higgins and Burns [13] and Alexander [2] provide a good general overview of the degradation of some of these compounds. Several more detailed reviews in papers and books are available [14,15,16, 17,18,19,20,21,22,23,24,25].

It is beyond the scope of this chapter to discuss in depth the specific pathways for the biotransformation of specific compounds by specific organisms. However, most biological transformations fit into a relatively small number of categories. Biological reactivity has been categorized both according to the general type of chemical or functional group involved, e.g., aromatic, alkyl, etc. [13,19], as well as by the general type of reactions occurring, e.g., oxidation, reduction, etc. [16,18,24].

Some typical reactions of some chemicals and the resulting products are given below. The reactions are not balanced. They do not show reactants which are either supplied from the environment (e.g., molecular oxygen) or by the organism. Some listed reactions may fit into more than one category. It must be stressed that the feasibility of any reaction depends on many environmental factors in addition to the specific organism present.

Some typical biologically mediated reactions are as follows:

1. Oxidation of Alkyl Compounds

$$RCH_3 \rightarrow RCH_2OH \rightarrow RCO_2H$$

2. Oxidative Dealkylation

$$ROCH_3 \rightarrow ROH + HCHO$$

3. Decarboxylation

$$RCOOH \rightarrow RH + CO_2$$

4. Aromatic Hydroxylation

$$Ar \rightarrow ArOH$$

5. Ring Cleavage

$$Ar(OH)_2 \rightarrow CHOCHCHCHCOHCOOH$$

6. β-Oxidation

$$CH_3CH_2CH_3COOH \rightarrow CH_3COOH + CH_3COOH$$

7. Epoxidation

$$RC = CR \rightarrow R\ C\text{-}C\ R \text{ (with O bridging the two C atoms)}$$

8. Oxidation of Sulfur

$$R_2S \rightarrow R_2SO \rightarrow R_2SO_2$$

9. Oxidation of an Amino Group

$$RNH_2 \rightarrow RNO_2$$

10. Hydrolytic Dehalogenation

$$RCHClCH_3 \rightarrow RCHOHCH_3 + Cl^-$$

11. Reductive Dehalogenation

$$RCCl_2R \rightarrow RCHClR + Cl^-$$

12. Dehydrohalogenation

$$RCH_2CHClCH_3 \rightarrow RHC = CHCH_3$$

13. Nitro-Reduction

$$RNO_2 \rightarrow RNH_2$$

14. Hydrolysis

$$RCH_2CN \rightarrow RCHONH_3$$

$$RC(O)OR \rightarrow RC(O)OH + HOR$$

$$RC(O)NR_2 \rightarrow RC(O)OH + HNR_2$$

KINETICS OF BIOTRANSFORMATION

Several rate expressions have been proposed to describe the kinetics of biotransformation in the environment. Most modeling work to date has focused on the rates of degradation in the aquatic system and not in soils. However, it is logical that similarities should exist between the two systems. Fundamental differences exist because a soil system is inherently a two-phase system with microorganisms primarily attached to surfaces [26].

Both population growth and soluble substrate utilization have been described by Monod kinetics [27] when the substrate is used to provide energy for growth and is not limited by any other needed substance. Several rate expressions commonly used to describe chemical degradation in the environment can be rationalized in terms of the equations proposed by Monod.

Monod's work resulted in the formulation of expressions relating the rate of microbial growth per unit volume, r_g $[MT^{-1}L^{-3}]$, and the rate of substrate utilization per unit volume, r_s $[MT^{-1} L^{-3}]$, to both substrate concentration and microbial biomass (equations 9-1 and 9-2),

$$r_g = \frac{\mu_m X S}{K_s + S} \tag{9-1}$$

$$r_s = \left(\frac{1}{Y}\right) \frac{\mu_m X S}{K_s + S} \tag{9-2}$$

where X is the microbial biomass per unit volume of liquid $[ML^{-3}]$, S is the substrate concentration $[ML^{-3}]$, and μ_m, k_s and Y are kinetic constants. μ_m is defined as the maximum growth rate constant $[T^{-1}]$, K_s is the half-saturation constant $[ML^{-3}]$, and Y is the true yield coefficient (a factor describing the efficiency of converting chemical mass into microbial mass). It should be noted that relationships (9-1) and (9-2) do not consider processes leading to the loss of biomass such as endogenous respiration. The term μ_m/Y is often replaced by k which is defined as the maximum rate of substrate utilization per unit mass of microorganisms

$$k = \frac{\mu_m}{Y} \tag{9-3}$$

to yield relationship (9-4)

$$r_s = \frac{k X S}{K_s + S} \tag{9-4}$$

Relationship (9-4) is a general biotransformation rate expression from which others can be rationalized. This rate expression indicates that the biotransformation rate is a function of properties at a particular <u>point</u> in the environmental system and is an algebraic expression not a differential expression. An expression of the form

$$\frac{dS}{dt} = \frac{-k\,X\,S}{K_s + S} \qquad (9\text{-}5)$$

which is commonly written as a rate expression reflects a mass balance on a batch (i.e., non-flow) system and does not in general describe the changing concentration of substrate in space and time within an environmental system.

Substrate removal following a form given in equation (9-4) is also sometimes referred to as following Michaelis-Menton kinetics after the work of Michaelis and Menton [28] who showed that the rate of substrate loss in an enzyme catalyzed reaction is of the form

$$r_{sE} = \frac{V_m\,S}{K_m + S} \qquad (9\text{-}6)$$

where r_{sE} is the rate of substrate reaction, K_m is the Michaelis-Menton constant, and V_m is the maximum reaction velocity which has a value directly proportional to the enzyme concentration, E

$$V_m = k'E \qquad (9\text{-}7)$$

where k' is a constant.

The similarity of equations (9-4) and (9-6) may be more than coincidental if the enzymatic nature of microbial degradation is such that the microorganisms may be treated as very small "packets" of enzymes. In any case, both Monod and Michaelis-Menton kinetics are frequently referred to when substrate utilization models are presented (even though the Monod model is based on empirical observation, and that of Michaelis-Menton is applicable only to enzyme kinetics).

Two limiting conditions may be encountered in the use of equation (9-4). At high substrate concentrations where S is much greater than K_s, the rate of substrate utilization becomes zero order in substrate

$$r_s = kX \qquad (9\text{-}8)$$

At low concentrations where K_s is much greater than S, the rate is first order in both substrate and biomass and second order overall

$$r_s = \frac{k}{K_s} XS = k_2 XS \qquad (9\text{-}9)$$

where $k_2 = k/K_s$ is a second order constant.

Further simplifications can be made if it is assumed that the microbial concentration X is constant with time, such as in the case in which a dynamic population equilibrium is established. New "constants" which depend on microbial concentration can be defined

$$k_X = kX \tag{9-10}$$

$$k_1 = k_2X \tag{9-11}$$

and substituted into equations (9-4), (9-8) and (9-9). The "exact" Monod expression then becomes

$$r_S = \frac{k_X\ S}{K_S + S} \tag{9-12}$$

At high substrate concentration the rate is a "constant"

$$r_S = k_X \tag{9-13}$$

and at low substrate concentration the rate is first order expression.

$$r_S = k_1S \tag{9-14}$$

Because the constant k_1 depends on microbial concentration, relationship (9-14) is sometimes referred to as a psuedo-first order relationship. A summary of all Monod based rate expressions is given in Table 9-1.

The direct applicability to soil systems of any rate expression based on a rigorous interpretation of Monod or Michaelis-Menton kinetics would not be expected since the microbial population in soil is diverse. These organisms may derive energy from a large number of substrates some of which are at a concentration too low

Table 9-1. Monod Based Rate Expressions.

Assumptions	Any X,S	High S $S > K_S$	Low S $S < K_S$
Any X,S	$r_S = \frac{k\ S\ X}{K_S + S}$	$r_S = kX$	$r_S = k_2\ SX$
X = constant	$r_S = \frac{k_XS}{K_S + S}$	$r_S = k_X$	$r_S = k_1\ S$

to support growth. Monod kinetics also does not consider the availability of substrates which may be limited by adsorption or mass transfer.

Nevertheless, rate expressions having similar form to those based on Monod kinetics have been used to describe the biotransformation of chemicals in the environment. While these expressions are not on a firm theoretical basis, they are logical. Logic dictates that the rate of biotransformation be a function of both substrate concentration and of the actively-degrading microbial population; it should approach zero as either X or S approaches zero; and it should increase as X or S increases at least over some range in concentration. Such a principle can be applied to both soluble and adsorbed chemical concentration as well as to attached and dispersed microorganisms.

Biotransformation rate expressions specifically applied to soils have generally not included microbial activity as a separate parameter. Two general rate expression types have been considered [16,29]: A power rate law

$$r_s = kS^n \tag{9-15}$$

where k is a constant, n is the reaction order, and S is the chemical concentration, and a "hyperbolic" expression

$$r_s = \frac{c_1 S}{c_2 + S} \tag{9-16}$$

where c_1 and c_2 are constants. Both expressions have been used to model overall chemical disappearance without regard to the relative importance of abiotic and biotic factors. In other words, they are general expressions describing apparent chemical loss, not necessarily general biotransformation rate expressions. Goring et al. [16], compared the use of these rate expressions for describing various degradation processes. The hyperbolic model is consistent with Monod kinetics, yielding a simple first order degradation rate expression at low chemical concentrations

$$r_s = k_1 S \tag{9-17}$$

where $k_1 = c_1/c_2$, and a zero order expression at high chemical concentrations.

$$r_s = c_1 \tag{9-18}$$

Rationalization of the hyperbolic expression with Monod kinetics requires the constant c_1 to be a function of microbial concentration which implies that expression (9-17) can be rewritten as a second order expression

$$r_S = k_2 S X \tag{9-19}$$

where X is microbial concentration and $k_2 = k_1/X$.

More complex Monod type expressions while potentially describing the rate of degradation more accurately are computationally more complex and require difficult-to-measure parameters. Furthermore, the use of more simple rate expressions may be rationalized at low chemical concentrations where a first order dependence on chemical concentration may be reasonable [30]. The biotransformation of several pesticides in soils has been found to follow first order kinetics [31,32]. However, more complex rate expressions have been used to describe degradation rates. McCarty et al. [12], have developed a biofilm model incorporating the more complex Monod type expression (9-4) (including microbial concentration as a parameter) with its full complement of constants and have suggested that the model may be applicable in predicting the fate of chemicals in the subsurface environment.

The use of the power-rate model cannot be rationalized in terms of Monod kinetics unless n equals zero or one (which corresponds to the high and low substrate concentration assumptions used in the Monod formulation). Goring et al. [16], have suggested the use of the power rate model where the goal is to develop an empirical relationship to fit degradation rate data without regard to the nature of the degradation mechanism. This level of empiricism should be avoided when using a fate model which presumably accounts for all important phenomena as separate components.

FACTORS INFLUENCING BIOTRANSFORMATION

A variety of environmental and chemical factors limit the biotransformation of chemicals in the environment [1].

Inaccessibility of the Chemical

The compound may be held in a microenvironment which precludes microbial attack, e.g., it may be sorbed or entrapped in a manner which prevents the organism or its enzyme from reaching the substrate.

Absence of Some Factor Essential for Growth

For example, no growth would occur in the absence of water, nitrogen, or phosphorus.

Toxicity of the Environment

This could be the result of biologically generated inhibitors, extremes in environmental temperature or pH, high salt concentration, high toxic levels of the chemical itself, or other inhibitory conditions.

Inactivation of the Requisite Enzymes

Enzyme activity may be lost by adsorption or inhibited by other substances (e.g., phenolic and polyaromatic substrates or products).

A Structural Characteristic of the Molecule

Some functional characteristics prohibit the formation of an enzyme-substrate complex.

Inability of the Community of Microorganisms to Metabolize the Compound Because of Some Physiological Inadequacy

An enzyme capable of degrading the compound may not exist, or the compound may not be able to penetrate into the cells in which the appropriate enzymes exist.

AVAILABILITY OF THE CHEMICAL

Availability of the chemical is discussed separately from major environmental factors because of its general importance. Both macroscopic and microscopic processes affect chemical availability. At the macroscopic level, microorganisms may be distributed in a very patchy manner precluding significant biotransformation where their numbers are relatively small. At the microscopic level, phenomena occur which cause chemicals to partition themselves between the liquid and solid phases and may play a major role in determining the overall rate of biotransformation by determining chemical availability. Because of the relatively poor understanding of the causes of this partitioning, the more general term sorption will be used in this discussion instead of the frequently used term, adsorption.

Upon sorption, biotransformation rates are frequently drastically changed. Both retardation and acceleration effects have been observed [33,34,35]. In most cases, sorption appears to reduce degradation rates. For example, Steen et al. [34], demonstrated that some chemicals were not available to aquatic microorganisms when sorbed onto sediments. However, growth of bacteria on surfaces to which substrates can sorb (and concentrate) has been observed in solutions too dilute to support a dispersed growth [35,36].

The effects of sorption on biodegradation may have several possible causes [37]:

1. A sorbed chemical may not be attacked by an enzyme or microorganism because of a physical barrier or inability of the enzyme to form a substrate-enzyme complex with the adsorbed chemical (rate decreases).

2. The chemical may not be sorbed in close enough proximity to a microorganism or enzyme so that interaction is delayed until the substrate flux brings the reactants together (rate decreases).

3. The sorbed chemical may not be concentrated in an area where proliferation of microbes can occur (rate decreases).

The effect of sorption on the mathematical expression of the degradation rate can be handled several ways. The most general approach would be to consider the biodegradation of the sorbed chemical as a separate rate term independent of the rate of soluble chemical degradation. For example, if chemical loss was assumed to be first order, then the rate of sorbed chemical loss per mass of soil (r_A) and the rate of soluble chemical loss per volume of liquid (r_W) could be expressed as

$$r_A = k_A S_A \tag{9-20}$$

$$r_W = k_W S_W \tag{9-21}$$

where k_A and k_W are first order biodegradation rate constants $[T^{-1}]$ specific to each phase, and S_A and S_W are the sorbed and soluble chemical concentration with units of mass sorbed per mass of soil and mass per liquid volume, respectively. The adsorbed concentration and rate could also be expressed per volume of soil by multiplying r_A by soil density.

The rate of total chemical loss can be easily related to the rates of loss from each phase for first order kinetics if rapid equilibrium is assumed. The rate of total chemical loss r_S can be expressed several ways: relative to soil water volume, to soil mass, or to soil volume. Each yields a first order expression of the same form,

$$r_S = k_T S_T \tag{9-22}$$

where k_T is a first order rate constant and S_T is the total chemical concentration. The rate constant k_T is a function of the partition coefficient K_D $[L^3m^{-1}]$, the ratio of soil mass to soil water volume, M, and the phase specific biotransformation rate constants, k_A and k_W according to

$$k_T = f_1 k_W + f_2 k_A \tag{9-23}$$

where f_1 and f_2 are the fraction of total chemical mass in the water and on the soil, respectively, and given by

$$f_1 = \frac{1}{1 + K_pM} \quad (9\text{-}24)$$

$$f_2 = \frac{K_pM}{1 + K_pM} \quad (9\text{-}25)$$

It should be stressed that while equations (9-20) and (9-21) may be regarded as "fundamental" biotransformation rate expressions, equation (9-22) is not, since it incorporates characteristics of the soil system which are independent of biotransformation. This should be kept in mind since biotransformation studies frequently measure a first order biotransformation constant that is in actuality not a single fundamental constant. It would clearly be in error to use a first order rate constant measured from simple observations of the rate of total chemical loss as a fundamental biotransformation rate constant in a model which also separately incorporates the effect of sorption on the rate of biotransformation.

Mass transfer limitations could also govern degradation rate by controlling the chemical concentration actually available to microorganisms. Mass transfer effects can be incorporated into the overall chemical mass balance. McCarty et al. [12], incorporated mass transfer limitations in a biofilm model which was suggested as being applicable to modeling of the subsurface environment.

MAJOR FACTORS AFFECTING BIODEGRADATION

Factors that influence the rate of biodegradation are listed in Table 9-2. These variables fall into three general categories: (1) those that affect substrate availability, (2) those that directly affect the microbial population size, composition, and activity (e.g., population interactions), and (3) those that directly control the degradation rate itself (e.g., temperature).

Most factors are not independent but are highly interrelated. For example, pH may affect both the availability of a substrate as well as the composition of the microbial community. The extent of the interrelation of factors frequently leads to overall effects on

Table 9-2. Major Environmental Factors Affecting Biotransformation.

1. pH	6. Oxygen
2. Temperature	7. Nutrients
3. Water content	8. Nature of microbial population
4. Carbon content	9. Acclimation
5. Clay content	10. Concentration

degradation that are not always the same since some factors may act to increase the rate of degradation while others may decrease it. While some generalizations regarding the net effect of each factor can be made, the literature is full of exceptions.

pH

Microbial activity is affected by the hydrogen ion concentration. Microorganisms can grow in a limited range of pH values. Optimum growth may occur at different pH values for different organisms. Most soil bacteria grow optimally near neutral conditions (pH 6.5 to 8.5). Actinomycetes are less tolerant of low pH as a group and generally are not found below pH 5; but they do grow relatively better than bacteria under alkaline conditions. Fungi can develop over a wide range of pH that generally spans from highly acidic pH (as low as 2 to 3) to alkaline conditions although individual species may show a lesser tolerance [2]. Therefore, as a soil becomes more acidic, the proportion of fungi to bacteria generally increases. In addition, individual enzymes produced by a specific microbe may be affected by pH and therefore may affect reaction rates with substrates.

Hydrogen ion concentration may affect the availability of substrate by influencing the amount of chemical sorbed. Hydrogen ions may compete for sorption sites thereby decreasing the amount sorbed. If the chemical ionizes, lowering the pH may either increase or decrease adsorption depending on whether the protonated chemical or nonprotonated chemical is more readily adsorbed [37]. Enzymatic activity may also favor one form of an ionizable chemical.

Temperature

Microbial activity is generally stimulated by temperature increases within the range tolerated by the microorganisms, and some species tend to dominate within certain temperature ranges [38]. Microorganisms may be classified into three major groups depending on the temperature range for optimum growth rate: mesophiles (optimum activity between 25 and 35°C), psychrophiles (optimum activity below 20°C), and thermophiles (optimum activity between 45 and 65°C). Many species of microorganism may have optimum growth over narrower or wider ranges.

Increasing rates of degradation with increasing temperature are a consequence of increased chemical reaction rates. This effect is sometimes expressed in terms of a Q_{10} value which is the ratio of the rate at a temperature 10 degrees higher than a reference temperature to that of the rate at the reference temperature. For example, a reaction characterized by a Q_{10} of 2.0 would double in rate for a 10 degree increase in temperature. Q_{10} values may be interpreted as reflecting a change in rate constants used in degradation rate expressions.

Other empirical relationships have been used to calculate biodegradation rate constants such as the use of empirical factors correlating temperature effects on rate constants for specific reactions

$$k_2 = k_1\theta^{(T_2-T_1)} \tag{9-26}$$

where k_1 is the rate constant at temperature T_1, k_2 is the rate constant at T_2, and θ is an empirical factor which depends on temperature [39].

Smith and Walker [40] developed an empirical expression relating an overall first order pesticide degradation rate constant to both soil temperature and soil-water moisture content, M

$$K_1 = \alpha MT + \beta \tag{9-27}$$

when α and β are constants which depend on specific pesticide and soil type. However, their equation was developed to correlate observed losses in pesticide without discerning between biological and nonbiological transformations.

Rate data have also been correlated using the Arrhenius equation

$$k = A \exp(-E/RT) \tag{9-28}$$

where k is the reaction rate constant, A is a preexponential constant, R is the gas constant, E is the activation energy (a constant), and T is the absolute temperature [41]. While the Arrhenius equation may be useful, it should be kept in mind that it is applicable to only single rate limiting chemical reactions and not generally to reactions that may involve several consecutive steps and that are common in microbial systems. Relationship (9-26) may be derived from consideration of relationship (9-28).

Adsorption processes are exothermic, and, therefore, an increase in temperature is expected to decrease the amount of chemical sorbed. This may increase the concentration which is available for degradation and hence may contribute to increased degradation. In general, decreased sorption and increased microbial activity associated with higher temperatures usually enhance pesticide degradation [42]. However, exceptions have been noted in which increased temperature has increased sorption possibly as a result of structural change in the adsorbing surface [37]. This could somewhat offset any increase in microbial activity. The overall effect would depend on the relative rates of biotransformation of sorbed and dissolved chemical.

Water Content

Moisture is required for microbial growth, but an oversupply can reduce gas exchange and limit oxygen which is depleted by microbial metabolism; thus, an anaerobic environment is created. This is because the rate of diffusion of oxygen in water is slow. An optimum level for aerobic organisms is about 50-75 percent of the soil moisture-holding capacity [2]. Bacterial and fungi populations have been closely correlated with moisture content [2].

Walker [43] correlated herbicide half-lives to soil moisture content, M using

$$t_{1/2} = AM^{-B} \tag{9-29}$$

where A and B are constants.

Moisture content affects soluble chemical concentration. Decreasing moisture causes increased sorption of chemical which may affect the degradation rate. Increased moisture could result in lower chemical concentrations in the aqueous phase and decreased biotransformation rates. Increased moisture could also dilute any potentially toxic chemical and thereby could possibly increase its biotransformation rate.

Carbon Content

Soil organic carbon content is due to organic matter consisting of an amorphous colloidal fraction or humus, in a macro- or microscopically definable component made up of plant and animal material at various stages of decay [37]. Organic carbon and associated nutrients are the major constituents of microbial food. Community size in mineral soils has been directly related to organic matter content [2].

Organic matter strongly sorbs many chemicals, and both persistence and sorption have been correlated with organic content [44, 45]. However, while sorption may tend to decrease the rate of degradation, the potential increase in microbial activity may counter this effect. Verma et al. [46], have also suggested that soluble humic polymers can act as stabilizing agents making compounds less resistant to biodegradation. Sequestering of a chemical could also make it more resistant to attack.

Additions of supplemental carbon such as sucrose or nutrient broth to soils have resulted in both increased [47,48] and decreased rates of biodegradation of a specific organic chemical [49,50]. A decreased rate may result from avoidance of the specific organic chemical because of a greater ease of utilization of the added carbon. An increased rate may be a result of increased microbial activity and cometabolism. However, addition of readily utilizable carbon sources generally increases degradation rates. These observations may have significant bearing on the biodegradability of chemical mixtures, particularly if one chemical is in great excess of another. Cometabolism and secondary substrate utilization may occur during the degradation of both the added as well as the naturally occurring carbon.

Clay Content

Clay contributes to the mineral fraction of soil. Many chemicals, including enzymes, sorb to clay. Clays are cation exchangers; so cationic chemicals may be readily sorbed.

The quantity and type of clay is important in determining the effect on degradation. Weber and Coble [33] added montmorillonite and kaolinite clays to nutrient solutions containing diquat. Montmorillonite sorbed the diquat, and biodegradation was reduced in direct proportion to the amount of clay added. Kaolin did not sorb diquat, and biodegradation was not affected by the quantity of clay added. Clay may also affect the oxygen content in a soil because a clay soil tends to retain a higher moisture content (which in turn may restrict oxygen diffusion).

Oxygen

Oxygen is needed as the terminal electron acceptor for some microorganisms, and the presence or absence of oxygen determines the specific species as well as the general type of organism present. Filamentous fungi and actinomycetes are, as a group, strict aerobes. Common soil bacteria include aerobes, strict anaerobes, and facultative aerobes.

Some compounds are transformed by microorganisms only under aerobic conditions while some are transformed under either aerobic or anaerobic conditions, or not at all [51,52,53]. The presence of oxygen not only may affect the biotransformation mechanism (and rate) but also the products formed. Pesticide mineralization rates are considerably lower, and substantial amounts of intermediate metabolites may accumulate under anaerobic conditions [42].

Soil environments are rarely totally aerobic. Soils tend to be more anoxic with depth because oxygen is consumed rapidly relative to the rate it diffuses in from the surface. Anaerobic microenvironments may exist even in well-aerated soil [2].

Nutrients

Nutrient is a general term referring to any substance required for growth. Nutrients supply elements which become part of the protoplasm (biomass), produce energy for cell growth and biosynthetic reactions, and serve as electron acceptors. The general nutritional requirements of microorganisms have been previously discussed in this chapter and will not be repeated here.

The consequence of a nutrient limitation may be inhibition of microbial proliferation or the inhibition of microbial respiration. The effect would be to reduce the rate of biodegradation. The availability of carbon, phosphorus, and nitrogen frequently affects microbial growth because these are the major constituents of protoplasm. These nutrients may be supplied by the chemical being degraded or by the environment.

Nature of the Microbial Population

The degradation of chemicals is characterized by microbial specificity, i.e., not all chemicals are degraded by all organisms. For example, cellulose degraders are predominantly fungal [37], and many hydrocarbons support few microbial species [54]. Differences also exist in the degradative pathways and rates exhibited by different species of microorganisms.

Population size and spatial distribution are important factors. Microhabitats may differ greatly in numbers of microorganisms, possibly because of the distribution of organic matter or other environmental factors. It is axiomatic that an increase in an active degrading population should increase the overall rate of degradation.

Interactions among species may indirectly affect biodegradation rates. Competition and predation affect microbial composition. Metabolic activity of one species may affect that of others through the production of cofactors or metabolic intermediates that can be used by another organism. Chemicals may be degraded sequentially by a mixture of species. Focht [55] showed that two microbes in liquid culture were capable of _together_ metabolizing DDT, but no single species has been discovered which can utilize DDT as a sole carbon and energy source. Healy et al. [56], have shown that a consortia of bacteria were responsible for the anaerobic transformation of ferulic and benzoic acids.

Acclimation

The degradation of a chemical frequently does not occur at an appreciable rate immediately after its introduction to an environment. This "lag" period in which the organism and system become acclimated may have several causes. Required enzymes may have to be induced, a change in the environment may allow for preferential selection of the degrading organism, or the added chemical may gradually increase the population of existing microbes capable of degrading the chemical [57].

The significance of prior acclimation, particularly for more complex compounds, is well known. For example, Whiteside and Alexander [58] demonstrated that 2,4-D was degraded in soil by microorganisms after a lag period. Furthermore, a second application was degraded more rapidly without a lag period. Presumably the required enzymes were still present after the first application had been degraded.

Concentration

High concentrations of chemicals may be bactericidal or bacteriostatic, thereby reducing the rate of degradation. Many pesticides have been shown to be more slowly degraded when applied to soils at high concentrations [59,60]. On the other hand, low concentrations may not be sufficient to initiate enzyme induction or support microbial growth.

ESTIMATION OF RATE PARAMETERS

Field vs. Laboratory Measurements

Evaluation of biotransformation rate constants under field conditions is difficult because of limitations imposed by concurrent abiotic processes (discussed in Limitations Section). In most field studies, the area is divided into plots, and the chemical is applied and appropriately incorporated to the desired depth. Inability to properly control the myriad of variables can result in meaningless numbers. Effective soil sterilization under field conditions is particularly difficult to achieve, if not impossible.

Laboratory tests using soil samples are required if the role of abiotic factors in degradation is to be assessed. A sterile control is often used along with a non-sterile soil sample to distinguish between abiotic and biotic factors. The soils may be incubated under either aerobic or anaerobic conditions. Anaerobic conditions can be maintained by flushing with nitrogen, nitrogen-carbon dioxide mixtures, or other inert gas. Aerobic conditions can be maintained by flushing with air or by leaving the sample open to the atmosphere. Redox potentials have been measured as an indicator of the relative oxidation state of a soil [61,62] under field conditions and have been suggested as a basis for classifying the oxygen levels in soils [63]. Flooded conditions are easily simulated in the laboratory. Wilson and Noonan [64] have recently discussed the use of laboratory microcosms to evaluate the transformation of chemicals in the subsurface environment. Methodologies for field and laboratory testing are available [65]. Methods used in sampling the subsurface environment and obtaining a representative soil sample are available [66,67,68].

Measurement of Test Chemical Disappearance

No attempt is usually made to differentiate between biotransformation of a sorbed chemical and that found in true solution. Measured biotransformation rate constants are therefore usually based on total chemical concentration in the soil and include sorption effects. The importance of phase possibly could be resolved through experiments utilizing various soil to water ratios and differentiation between dissolved and sorbed chemical.

Biotransformation of a chemical under field conditions has generally been measured as the rate of disappearance of solvent-extractable parent compound. Therefore, the extent of transformation of the structure of the parent chemical is not considered, and accumulation of metabolic intermediates cannot be ruled out. Chemical extraction and specific analytical methodologies for quantification of chemical concentration are discussed elsewhere [65,69].

Mineralization rates which indicate complete destruction of the parent compound can be measured under laboratory conditions by using ^{14}C labeled chemicals and collecting evolved $^{14}CO_2$ in a solution such as KOH. The rate of mineralization can be determined from the rate at which $^{14}CO_2$ is produced. Biotransformation rate constants based on $^{14}CO_2$ formation are expected to be smaller than those based on solvent extraction of the parent compound [42]. However, abiotic reactions resulting in decarboxylation of the chemical could cause an erroneously high apparent biotransformation rate. Inability to extract all the parent compound would also lead to misleadingly high rates.

Biotransformation Rate Expressions and Constants

Biotransformation rate expressions in soil systems have been expressed primarily as a first order rate relationship. Data analysis is simplified since only chemical concentrations need to be measured. In principle, first order rate constants can be obtained from batch experiments from the slope of the relationship

$$\ln S/S_0 = -k_1 t \qquad (9\text{-}30)$$

where S is the substrate concentration at time t, S_0 is the initial substrate concentration, and k_1 is the first order rate constant. In addition, substrate half-life $t_{1/2}$ is easily related to the first order rate constant by

$$k_1 = \ln(2)/t_{1/2} = 0.693/t_{1/2} \qquad (9\text{-}31)$$

This approach is frequently taken even if it is apparent that the time dependence of chemical disappearance is more complex. Ou et al. [70], and Rao and Davidson [42] have compiled an extensive list of first order transformation rate constants and half-lives for selected pesticides based on laboratory or field studies under aerobic or anaerobic conditions. It should be noted that in most studies, experimental conditions were not controlled to eliminate abiotic transformation pathways so that the constants obtained cannot necessarily be attributed to microbial action only.

Second order rate constants can be developed if, in addition to chemical concentration, microbial biomass or some surrogate parameter is measured. Second order rate constants are frequently expressed relative to biomass concentration or microbial numbers. Evaluation of a second order constant would require experiments to be conducted over a range of microbial biomass. Second order constants could then be related to pseudo-first order constants obtained at a constant microbial concentration X by

$$k_1 = k_2 X \quad (9\text{-}32)$$

from which k_2 can be obtained as the slope of a k_1 vs. X plot.

Measurement of Microbial Numbers and Biomass

Determination of second order rate constants requires that some population parameter directly proportional to chemical biotransformation rate be measured and converted to either equivalent biomass or numbers. Because the actively degrading population is impossible to measure accurately, estimates of total biomass and numbers are made.

Proper attention must be given to sampling, storage, and extraction methods which are used to separate microorganisms from the soil matrix. Microbiological sampling methods have been discussed by Gilmore [66], Atlas and Bartha [68], McNabb and Mallard [71], Williams and Gray [72], and Board and Lovelock [73]. Many approaches have been used to estimate microbial numbers and biomass in the subsurface environment. Leach [74] has recently provided an excellent review of commonly used methods as well as those that are of a more experimental nature. Methodologies reviewed included bioassays, measurement of growth, staining and microscopy, metabolic and physiological responses, measurement of enzyme activity, and the determination of cellular components [74]. Methods vary greatly in precision, ease, and lower detection limit of microbial numbers. Several of the more commonly used methods are briefly discussed here.

Microbial Numbers

Microbial numbers are frequently enumerated by plating methods and by direct microscopic evaluation. Alexander [2] has discussed general methods applicable in enumerating microorganisms.

Plate count methods utilize a soil sample of known size diluted with water and agitated to disperse microorganisms. Samples are withdrawn and either mixed with a cooled agar media (pour plate method) or spread onto a pre-dried agar plate (spread plate method) which is then incubated. Each colony formed is assumed to have originated from a single microorganism. Plating methods may rely on a general media on which a wide variety of heterotropic organisms are known to grow. Different media may be used to differentiate fungi from bacteria and actinomycetes.

Epifluorescence microscopy utilizing acridine-orange (AO) as a stain has been used to determine microbial numbers in soils [75] and subsurface samples [76]. Both living and dead cells are counted. Electron microscopy has also been utilized [77]. Many more bacteria are usually counted by direct epifluorescence than by conventional plate counts in aquatic samples, and this suggests that epifluorescence may be the better method in enumerating bacterial numbers in soils [78]. The results of cell counts can be converted to biomass by applying appropriate volume and density factors [79], or the numbers can be reported as direct counts.

Biomass

Measurement of the concentration of a chemical that is a cellular component can be used to estimate cell number and biomass if a relationship between them is assumed. Adenosine triphosphate (ATP) is present in all living organisms and can be extracted and measured by determining the amount of light produced by its reaction with added firefly luciferace. The use of ATP measurements as a measure of biomass has been discussed by Leach [80], Eiland [81], Jenkinson and Ladd [82], and Eiland and Nielsen [83]. Stevenson et al. [84]. summarized the advantages of using ATP as an indicator of microbial populations:

1. Sparse communities are detectable.
2. All physiological types of microbes are included.
3. Only living cells respond.
4. The method is quick.
5. The method is precise.

Problems with using ATP are associated with the extraction efficiency, presence of interfering substances, variability in the amount of ATP per unit biomass (which depends on the particular organism species and its nutritional state), and the instability of ATP unless the sample is properly preserved. However, even given these limitations, ATP measurements may be more reliable as a biomass indicator than those based on plating methods [85]. Webster et al. [76], have recently compared the use of acridine-orange staining and ATP measurements to determine cell numbers in subsurface samples. The measurement of chloroform extractable lipid phosphate is also a simple technique [86].

LIMITATIONS TO APPLYING DEGRADATION RATE EXPRESSIONS

There are a number of practical as well as theoretical limitations in the use of degradation rate expressions which may result in discrepancies between predicted and measured chemical concentrations. These limitations arise from measurement difficulties and assumptions made in arriving at a rate expression, and they are summarized in Table 9-3.

Table 9-3. Limitations in Applying Degradation Rate Expressions.

Measurement	Model
1. Abiotic vs. biotic processes	1. Specificity of constants
2. Microbial concentrations	2. Concentration dependence of constants
3. Bound chemical residues	3. Acclimation

Measurement Limitations

Abiotic vs. Biotic Processes

Biological transformations may be difficult to distinguish from abiotic processes without carefully controlled experiments. Abiotic processes include strictly chemical reactions, photochemical reactions, and physical processes such as leaching and volatilization. The effect of volatilization has been previously discussed. Chemicals may undergo various transformations such as hydrolysis, oxidation, and reduction in the absence of microorganisms or their enzymes [19,87]. The products of nonbiological degradation are frequently identical to those resulting from biological reactions [37].

A common approach to resolving biotic and abiotic processes is to inhibit microbial growth and physiological processes in the soil under laboratory conditions that minimize physical losses such as volatilization. Soil sterilization has been accomplished by chemical addition (e.g., sodium azide), heat treatment, and irradiation [6].

Some sterilization procedures such as heat treatment may alter physical and chemical properties of the soil which affect abiotic processes. Chemical treatment may inactivate microorganisms but have little effect on extracellular enzymes. Furthermore, microorganisms may be rendered incapable of true growth but still capable of respiration. Some methods of microbe inactivation may release enzymes which still function outside the cell. Presumably the absence of chemical transformation in a sterilized soil indicates the absence of abiotic processes in non-sterilized soil. Furthermore, transformations occurring in a sterilized soil are assumed totally attributable to abiotic processes. However, neither assumption may be true.

Microbial Concentration

Application of a degradation rate expression (e.g., a second order relationship) may require measurement of microbial biomass or numbers. Unfortunately, microbial biomass and numbers may not always correlate well with observed degradation rates [88,89]. This may have several causes beyond those associated with imprecise or inappropriate methodology. The total microbial population is not necessarily responsible for chemical degradation, and the fraction responsible may change with location. Ideally, only the "active" population should be measured. The action of extra-cellular enzymes may also account for a significant fraction of chemical transformation. Fundamental differences in metabolic activity could exist as a result of environmental changes.

Estimates of microbial numbers and biomass depend on the method used to obtain them. For example, plating methods may involve the use of a media on which some microbial species will not grow. Several microorganisms may be clumped together and appear as a single colony [2], or incubation times may not be long enough for some organisms to grow. Estimates of fungi may also be in error because colonies appearing on agar may be derived from a spore or a fragment of vegetative mycelium. Direct microscopic determination, such as epifluorescence microscopy, cannot discern viable from non-viable cells. Chemical methods may also suffer from this problem as well as having a variable ratio of measured chemical to biomass [90].

Bound Chemical Residues

Chemicals or their transformation products added to soil may react with material in the soil and become "bound" [14]. Kaufman et al. [49], defines a bound pesticide residue as "that unextractable and chemically unidentifiable pesticide residue remaining in fulvic acid, humic acid, and humic fractions after exhaustive sequential extraction with non-polar organic and polar solvents." Therefore, inability to recover an added chemical from soil may not be due entirely to biodegradation.

Model Limitations

All of the degradation rate expressions are empirical albeit logical and capable of being rationalized in terms of Monod kinetics or Michaelis-Menton kinetics. As a consequence, use of the results at conditions not used to develop or evaluate the rate expressions must be done with caution.

Specificity of Constants

Field and laboratory measurements of rate constants will only be comparable when experimental conditions (especially temperature, moisture content, soil properties, nutrient concentrations, oxygen concentration, and chemical concentration) are similar. Use of such data requires appropriate caution because of unknown factors. In general, transformation rates under field conditions appear greater than those under laboratory incubation studies because of multiple processes occurring in the field [42,91]. In particular, first order constants are inherently more site specific than second order constants since a first order constant does not expressly consider changes in microbial activity. Constants based on the rate of disappearance of total chemical concentration may be a function of the degree of chemical sorption.

Concentration Dependence of Constants

Degradation constants are assumed to be independent of concentration. However, this may not be true. For example, Boethling and Alexander [92,93] have shown that the Michaelis-Menton constants determined at high concentrations may not be consistent with the observed degradation rates at low chemical concentrations.

Additionally, use of an inadequate rate expression could give rise to an apparent concentration dependence. For example, a first order dependence on substrate is not expected at high substrate concentration. Evaluation of first order constants at both low and high concentrations could yield two different "constants." Rate constants may also appear to increase when evaluated at lower concentrations of a chemical if a chemical exhibits some inhibitory effect on growth at higher concentrations.

Acclimation

Rate constants will depend on the extent of acclimation of the microbial population to the test chemical. Microbial populations in the field may not be exposed to the chemical long enough for acclimation to occur. Hence, the observed degradation rates may be lower than those predicted from constants obtained from acclimated populations. Acclimation times have been observed to range from a few hours to a few months depending on the chemical structure and its concentration.

REFERENCES

1. Alexander, M. "Biodegradation: Problems of Molecular Recalcitrance and Microbial Infallibility," Adv. Appl. Microb. 7:35-80 (1965).

2. Alexander, M. Introduction to Soil Microbiology, 2nd Ed. (New York: John Wiley & Sons, 1977).

3. Craigie, J. S., M. McLachlan, and G. H. N. Towers. "A Note on the Fission of an Aromatic Ring by Algae," Can. J. Bot. 43: 1589-1590 (1965).

4. Swisher, R. D. "Surfactant Biodegradation (New York: Marcel-Dekker, Inc., 1970).

5. Stotzky, G. "Activity, Ecology and Population Dynamics of Microorganisms in Soil," in Microbiol Ecology, A. I. Laskin and H. Lechevalier, Eds. (Cleveland, OH: CRC Press, 1974).

6. Skujins, J. J. "Enzymes in Soil," in Soil Biochemistry, A. D. McClaren and G. H. Peterson, Eds. (New York: Marcel-Dekker, Inc., 1967).

7. Getzin, L. W., and I. Rosefield. "Organophosphorous Insecticide Degradation by Heat-Labile Substances in Soil," J. Agr. Food Chem. 16:598-601 (1968).

8. Getzin, L. W., and I. Rosefield. "Partial Purification and Properties of a Soil Enzyme that Degrades the Insecticide Malathion," Bioch. Bioph. Acta. 235:422-53 (1971).

9. Horvath, R. S. "Microbial Co-Metabolism and the Degradation of Organic Compounds in Nature," Bacteriol. Rev. 36:146-155 (1972).

10. Perry, J. J. "Microbial Cooxidations Involving Hydrocarbons," Microb. Rev. 43:59-72 (1979).

11. Jacobson, S. N., N. C. O'Mara, and M. Alexander. "Evidence for Cometabolism in Sewage," Appl. Environ. Microb. 40:917-921 (1980).

12. McCarty, P. L., M. Reinhard, and B. E. Rittman. "Trace Organics in Ground Water," Environ. Sci. Tech. 15:40-51 (1981).

13. Higgins, I. J., and R. G. Burns. The Chemistry and Microbiology of Pollution (London, New York, San Francisco: (1940).

14. Hill, I. R. "Microbial Transformations of Pesticides," in Pesticide Microbiology, I. R. Hill and S. J. L. Wright, Eds. (London, New York: Academic Press, 1978).

15. Matsumura, F., and H. J. Benezet. "Microbial Degradation of Insecticides," in Pesticide Microbiology, I. R. Hill and S. J. L. Wright, Eds. (London, New York: Academic Press, 1978).

16. Goring, C. A. I., D. A. Laskowski, J. Hamaker, and R. W. Miekle. "Principles of Pesticide Degradation in Soil," in Environmental Dynamics of Pesticides, R. Hague and F. H. Freed, Eds. (Plenum Press: 1975).

17. Bollag, J. M. "Microbial Transformations of Pesticides," Adv. Appl. Microb. 18:75-130 (1974).

18. Kaufman, D. D., and P. C. Kearney. "Microbial Transformations in the Soil," in Herbicides, Physiology, Biochemistry, Ecology, Vol. 2, L. J. Audus, Ed. (London, New York: Academic Press, 1976).

19. Helling, C. J., P. C. Kearney, and M. Alexander. "Behavior of Pesticides in Soil," Adv. Agron. 23:147-240 (1971).

20. Woodcock, D. "Microbial Degradation of Fungicides, Fumigants and Nematocides," in Pesticide Microbiology, I. R. Hill and S. J. L. Wright, Eds. (London, New York: Academic Press, 1978).

21. Cripps, R. E., and T. R. Roberts. "Microbial Degradation of Herbicides," in Pesticide Microbiology, I. R. Hill and S. J. L. Wright Eds. (London/New York: Academic Press, 1978).

22. Laveglia, J., and P. A. Dahn. "Degradation of Organophosphorus and Carbamate Insecticides in the Soil and by Soil Microorganisms," Ann. Rev. Entmol. 22:483-513 (1977).

23. Bollag, J. M. "Biochemical Transformation by Soil Fungi," Crit. Rev. Microb. 2:35-58 (1972).

24. Crosby, D. G. "The Fate of Pesticides in Environment," Ann. Rev. Plant Physiol. 24:467-492 (1973).

25. Alexander, M. "Biodegradation of Chemicals of Environmental Concern," Science 211:132-138 (1981).

26. Marshall, K. C. Interfaces in Microbial Ecology (Cambridge, MA: Harvard Press, 1976).

27. Monod, J., 1949. "The Growth of Bacterial Cultures," Ann. Rev. Microb. 3:371-394 (1949).

28. Michaelis, L., and M. L. Menton. "Die Kihetik der Invertinwirkung," Biochemische Zietschrift. 49:333-369 (1913).

29. Hamaker, J. W. "Mathematical Predictions of Cumulative Levels of Pesticides in Soil," Adv. Chem. Series. 60:122 (1966).

30. Hamaker, J. W. "Decomposition: Quantitative Aspects," in Organic Chemicals in the Soil Environment, Vol. 1, C. A. I. Goring and J. W. Hamaker, Eds. (New York: Marcel-Dekker, Inc. 1972).

31. Meikle, R. W., C. R. Younston, R. T. Hedlund, C. A. I. Goring, J. W. Hamaker, and W. W. Addington. "Measurement and Prediction of Picloram Disappearance Rates from Soil," Weed Sci. 21:549-555 (1973).

32. Walker, W. W., and B. J. Stojanovic. "Microbial vs. Chemical Degradation of Malathion in Soil," J. Environ. Qual. 2:229-232 (1973).

33. Weber, J. B., and H. D. Coble. "Microbial Decomposition of Diquat Adsorbed on Montmorillonite and Kaolin Clays," J. Agr. Food Chem. 16:475-478 (1968).

34. Steen, W. C., D. F. Paris, and G. L. Baughman. "Effects of Sediment Sorption on Microbiol Degradation of Toxic Substances," Proceedings of the American Chemical Society Symposium on Processes Involving Contaminants and Sediments, Honolulu, HI (1979).

35. Kjellenberg, S., B. A. Humphrey, and K. C. Marshall. "Effect of Interfaces on Small Starved Marine Bacteria," Appl. Environ. Microb. 43:1166-1172 (1982).

36. Heukelekian, H., and A. Heller. "Relation Between Food Concentration and Surface for Bacterial Growth," J. Bacteriol. 40:547-558 (1940).

37. Burns, R. G. "Factors Affecting Pesticide Loss from Soil," in Soil Biochemistry, Vol. 4, E. A. Paul and A. D. McLaren, Eds. (New York: Marcel-Dekker, Inc., 1975).

38. Edwards, C. A. "Factors Affecting the Persistence of Insecticides in Soil," Soils and Fert. 27:451-454 (1964).

39. Grady, C. P., and H. C. Lim. Biological Wastewater Treatment: Theory and Application (New York: Marcel-Dekker, Inc., 1980).

40. Smith, A. E., and A. Walker. "A Quantitative Study of Asulam Persistence in Soils," Pest. Sci. 8:449-456 (1977).

41. Lee, R. F., and C. Ryan. "Microbial Degradation of Organochlorine Compounds in Estuarine Waters and Sediments," EPA-600/9-79-012, Proceedings of the Workshop on Microbial Degradation of Pollutants in Marine Environments, Washington, DC (1979).

42. Rao, P. S. C., and J. M. Davidson. "Estimation of Pesticide Retention and Transformation Parameters Required in Non-Point Source Pollution Models," in Environmental Impact of Non-point Source Pollution, M. R. Overcash and J. M. Davidson, Eds. (Arbor Arbor, MI: Ann Arbor Science Publications, 1980).

43. Walker, A. "The Degradation of Metazule in Soil. I. Effects of Soil Type, Soil Temperature and Soil Moisture Content," Pest. Sci. 9:326-332 (1978).

44. Upchurch, R. P., and W. C. Pierce. "The Leaching of Monuron from Lakeland Sand Soil. II. The Effect of Soil Temperature, Organic Matter, Soil Moisture and Herbicide," Weeds 6:24-33 (1958).

45. Hartley, G. S. "Physical Behavior in the Soils," in Herbicides: Physiology, Biochemistry, Ecology, L. J. Audus, Ed. (London: Academic Press, 1964).

46. Verma, L., J. P. Martin, and K. Haider. "Decomposition of Carbon-14-Labeled Proteins, Peptides, and Amino Acids, Free and Complexed with Humic Polymers," Soil Sci. Soc. Amer. Proc. 39:279-284 (1975).

47. McCormick, L. L., and A. E. Hiltbold. "Microbiological Decomposition of Atrazine an Diuron in Soil," Weeds 14:77-82 (1966).

48. Miyazaki, S., G. M. Boush, and F. Matsumura. "Metabolism of ^{14}-Cchlorapropylate by Rhodotorula gracilis," Appl. Microb. 18:972-976 (1969).

49. Kaufman, D. D., J. R. Plimmer, P. C. Kearney, J. Blake, and F. S. Guardia. "Chemical Versus Microbial Decomposition of Amitrole in Soil," Weed Sci. 16:266-272 (1968).

50. McClure, G. W. "Accelerated Degradation of Herbicides in Soil by the Application of Microbial Nutrient Broths," Contrib. Boyce Thomson Inst. 24:235-240 (1970).

51. Guenzi, W. D., and W. E. Beard. "Aerobic Conversion of DDT to DDD and Aerobic Stability of DDT in Soil," Soil Sci. Soc. Amer. Proc. 32:522-524 (1968).

52. Sethunathan, N., and I. C. MacRae. "Persistence and Biodegradation of Diazinon in Submerged Soils," J. Agr. Food Chem. 17:221-225 (1969).

53. Bower, E. J., B. E. Rittman, and P. L. McCarty. "Anaerobic Degradation of Halogenated 1- and 2-Carbon Organic Compounds," Environ. Sci. Techn. 15:596-599 (1981).

54. Brock, T. D. Biology of Microorganisms (Englewood Cliffs, NJ: Prentice-Hall, Inc., 1970).

55. Focht, D. D. "Microbial Degradation of DDT Metabolites to Carbon Dioxide Water and Chloride," Bull. Environ. Contam. Toxic. 7:52-56 (1972).

56. Healy, J. B., L. Y. Young, and M. Reinhard. "Methanogenic Decomposition of Ferulic Acid, a Model Lignin Derivative," Appl. Env. Microb. 39:436-444 (1980).

57. Brock, T. D. Principles of Microbial Ecology (Englewood Cliffs, NJ: Prentice-Hall, Inc., 1966).

58. Whiteside, J. S., and M. Alexander. "Measurement of Microbiological Effects of Herbicides," Weeds 8:204-213 (1960).

59. Ahmed, N., and F. O. Morrison. "Longevity of Residues of Four Organophosphate Insecticides in Soil," Phytoprotection 53: 71-74 (1972).

60. Davidson, J. M., P. S. C. Rao, L. T. Ou, W. B. Wheeler, and D. F. Rothwell. "Adsorption Movement and Biological Degradation of Large Concentrations of Selected Pesticides in Soil," EPA-600/2-80-124 (U.S. EPA, 1980).

61. Quispel, A. "Measurement of the Oxidation-Reduction Potentials of Normal and Inundated Soils," Soil Sci. 63:265-275 (1949).

62. Bohn, H. L. "Redox Potentials," Soil Sci. 112:39-45 (1971).

63. Patrick, W. H., and J. R. Mahapatra. "Transformation and Availability to Rice of Nitrogen and Phosphorus in Waterlogged Soils," Advan. Agron. 20:323-359 (1968).

64. Wilson, J., and M. J. Noonan. "Microbial Activity in Model Aquifer Systems," in Groundwater Pollution Microbiology, G. Britton and C. P. Gerba, Eds. (New York: John Wiley & Sons, 1984).

65. Howard, P. H., J. Saxena, P. R. Durkin, and L. T. Ou. "Review and Evaluation of Available Techniques for Determining Persistence and Routes of Degradation of Chemical Substances in the Environment," EPA-560/5-75-006 (U.S. EPA, 1975).

66. Gilmore, A. E. "A Soil Sampling Tube for Soil Microbiology," Soil Sci. 87:95-99 (1959).

67. Wilson, L. G. "Monitoring in the Vadose Zone. A Review of Technical Elements and Methods," EPA-600/7-80-134 (Las Vegas, NV: U.S. EPA, 1980).

68. Atlas, R. M., and R. Bartha. Microbial Ecology (Addison-Wesley Publications, 1981).

69. Sherma, J. "Manual of Analytical Quality Control for Pesticides and Related Compounds," EPA-600/2-81-059 (Research Triangle Park, NC: U.S. EPA, 1981).

70. Ou, L. T., J. M. Davidson, P. S. C. Rao, and W. B. Wheeler. "Pesticide Transformations in Soils," in Retention and Transformation of Selected Pesticides and Phosphorus in Soil-Water Systems: A Critical Review, P. S. C. Rao and J. M. Davidson, Eds., EPA-600/3-82-060 (U.S. EPA, 1982).

71. McNabb, J. F., and G. E. Mallard. "Microbiological Sampling in the Assessment of Groundwater Pollution," in Groundwater Pollution Microbiology, G. Britton and C. P. Gerba, Eds. (Wiley-Interscience, 1984).

72. Williams, S. T., and T. R. G. Gray. in Sampling-Microbiological Monitoring of Environments, R. G. Board and D. W. Lovelock, Eds. (New York: Academic Press, 1973).

73. Board, R. G., and D. W. Lovelock. Sampling - Microbiological Monitoring of Environments (New York: Academic Press, 1973).

74. Leach, F. R. "Biochemical Indicators of Groundwater Pollution," in Groundwater Pollution Microbiology, G. Britton and C. P. Gerba, Eds. (Wiley-Interscience, 1984).

75. Trolldenier, G. "The Use of Epifluorescence Microscopy for Counting Soil Microorganisms," in Modern Methods in the Study of Microbial Ecology, T. Rosswall, Ed. (Bulletins from the Ecological Research Committee, Swedish Natural Science Research Council, Stockholm. 17:55-59, 1973).

76. Webster, J. J., G. J. Hampton, J. T. Wilson, W. C. Chiorse, and F. R. Leach. "Determination of Microbial Cell Numbers in Subsurface Samples," Ground Water 23:17-25 (1985).

77. Gray, T. R. G. "Stereoscan Electron Microscopy of Soil Microorganisms," Science 155:1668-1670 (1967).

78. Daley, R. J. "Direct Epifmorescence Enumeration of Native Aquatic Bacteria: Uses, Limitations, and Comparative Accuracy," in Native Aquatic Bacteria: Enumeration, Activity, and Ecology, J. W. Costerton and R. R. Colwell, Eds. (ASTM/STP 695:29, 1979).

79. Dale, N. G. "Bacteria in Intertidal Sediments - Factors Related to Their Distribution," Limnol. Ocean. 19:509-518 (1974).

80. Leach, F. R. "ATP Determination with Firefly Luciferase," J. Appl. Biochem. 3:473-517 (1981).

81. Eiland, F. "An Improved Method for Determination of Adenosine Triphosphate (ATP) in Soil," Soil Biol. Biochem. 11:31-35 (1979).

82. Jenkinson, D.S., and J. N. Ladd. in Soil Biochemistry, E. A. Paul and J. N. Ladd, Eds. (New York: Marcel-Dekker, Inc. 1981).

83. Eiland, F., and B. S. Nielsen. "Influence of Cation Content on Adenosine Triphosphate Determinations in Soil," Microb. Ecol. 5:129-137 (1979).

84. Stevenson, L. H., T. H. Chrzanowski, and C. W. Erkenbrecher. "The Adenosine Triphosphate Assay: Conceptions and Misconceptions," ASTM/STP 695:99-111 (1979).

85. Cavari, B. "ATP in Lake Kinneret: Indication of Microbial Biomass or of Phosphate Deficiency," Limnol. Ocean. 21:231-236 (1976).

86. White, D. C., W. N. Davis, J. S. Nickles, J. D. King, and R. J. Bobbie. "Determination of Sedimentary Microbial Biomass by Extractable Lipid Phosphate," Oecologo 40:51-62 (1979).

87. Crosby, D. G. "Non-Biological Degradation of Herbicides in the Soil," in Herbicides, Physiology, Biochemistry, Ecology, 2nd edition, L. J. Audus, Ed. (London: Academic Press, 1976).

88. Wright, R. T. "Natural Heterotrophic Activity in Estuarine and Coastal Waters," EPA-600/9-79-012, Proceedings of the Workshop on Microbial Degradation of Pollutants in the Marine Environment, Washington, DC (1979).

89. Nesbitt, J. J., and J. R. Watson. "Degradation of Herbicide 2,4-D in River Water - I. Description of the Study Area and Survey of Rate Performing Factors," Water Res. 14:1683-1688, (1980).

90. Karl, D. M. "Cellular Nucleotide Measurements and Applications in Microbial Ecology," Microb. Rev. 44:739-796 (1980).

91. Rao, P. S. C., and J. M. Davidson. "Retention and Transformation of Selected Pesticides and Phosphorus in Soil-Water Systems: A Critical Review," EPA-600/3-82-060 (U.S. EPA, 1982).

92. Boethling, R. S., and M. Alexander. "Microbial Degradation of Organic Compounds at Trace Levels," Environ. Sci. Technol. 13:889-911 (1979).

93. Boethling, R. S., and M. Alexander. "Effect of Concentration of Organic Chemicals on their Biodegradation by Natural Microbial Communities," Appl. Environ. Microb. 37:1211-1216 (1979).

CHAPTER 10

NONBIOLOGICAL TRANSFORMATION

R. L. Valentine
University of Iowa, Iowa City, Iowa

PROCESS DESCRIPTION

While all processes leading to structural changes in chemicals can be considered as occurring as the result of chemical reactions, these processes can be categorized as being either biological, chemical, or photochemical transformations. Biological transformations, i.e., transformations mediated by microorganisms, are discussed elsewhere. Chemical transformations of importance in the soil environment include hydrolysis and oxidation-reduction reactions (Redox reactions). Photochemical transformations can occur only in the presence of light and hence are expected to be important only at or very near the soil surface.

Hydrolysis

Chemical Hydrolysis refers to a general class of chemical reactions involving water which result in the net exchange of a hydroxyl group, OH^-, with some other group, X [1]. This occurs principally by cleavage of carbon-X and phosphorus-X bonds. The leaving group depends on the type of chemical. Common leaving groups include halides (Cl^-, Br^-), alcohols ($R-O^-$), amines ($R_1R_2N^-$), and sulfur (RS^-) and phosphorus containing moieties ($[RO]_2PO_2^-$). Several organic functional groups are particularly susceptible to hydrolysis [2]. These include alkylhalides, amides, amines, carbamates, carboxylic acid esters, epoxides, nitriles, phosphonic acid esters, phosphoric acid esters, sulfonic acid esters, and sulfuric acid esters. Many functional groups are not very susceptible to hydrolysis. These include saturated and unsaturated alkanes, aromatic hydrocarbons and amines, halogenated aromatics, aromatic nitro compounds, alcohols, ketones, glycols, phenols, ethers, carboxylic acids, sulfonic acids, and polycyclic and heterocyclic polycyclic aromatic hydrocarbons [2].

Hydrolysis reactions represent one class of reactions involving water. Several others may occur which include acid-base reactions, hydration of carbonyl compounds, elimination reactions, and addition to carbon-carbon double bonds. The first two reactions are considered reversible while the latter are not, and therefore could lead to permanent chemical transformation. A discussion of these is beyond the scope of this chapter.

Mabey and Mill [1] have discussed the hydrolysis of 12 classes of compounds of environmental importance. Many reactions yield a single product, however, several products may be possible if more than one site for attack exists as exemplified by the hydrolysis of malathion [3]. Several typical general reactions are presented.

Examples of Hydrolysis

a. $R - Cl + H_2O \rightarrow ROH + H^+ + Cl^-$

an alkyl halide → an alcohol

b. $R_1C(O)OR_2 + H_2O \rightarrow R_1C(O)OH + R_2OH$

an ester → a carboxylic acid + an alcohol

c. $R_1CH - CHR_2$ (bridged by O) $+ H_2O \rightarrow R_1CH(OH)CH(OH)R_2$

an epoxide → a diol

d. $ROC(O)NR_1R_2 + H_2O \rightarrow ROH + CO_2 + HNR_1R_2$

a carbamate → an alcohol + an amine

e. $RC(ON)R_1R_2 + H_2O \rightarrow RCOOH + R_1R_2NH$

an amide → a carboxylic acid + an amine

f. $RCH_2CN + H_2O \rightarrow RCH_2COOH + NH_3$

a nitrile → a carboxylic acid + ammonia

Mechanism and Kinetics

Hydrolysis is the result of a nucleophilic substitution reaction in which water or a hydroxide ion (a nucleophile) attacks electrophilic carbon or phosphorus and displaces a leaving group.

These nucleophilic substitution reactions occur by either a unimolecular (S_N1) or bimolecular (S_N2) mechanism. These general mechanisms are described in more detail elsewhere [4,5].

Aqueous hydrolysis has been characterized as potentially involving three distinct mechanistic processes. These are an acid catalyzed reaction, a neutral reaction, and a base promoted reaction. In pure water, a hydrogen ion (H^+) may catalyze the hydrolysis of a compound by reacting with it in such a way as to promote nucleophilic attack. Neutral hydrolysis processes involve attack by water alone. Hydroxide ion may also directly attack the electrophilic carbon or phosphorus to displace the leaving group.

In water, each process can give rise to a separate reaction term first order in chemical concentration, [S]. The general overall rate of hydrolysis (r_h) is the sum of these three terms:

$$r_h = k_H[H^+][S] + k_{H_2O}[H_2O][S] + k_{OH}[OH^-][S] \quad (10\text{-}1)$$

where k_H, k_{H_2O}, and k_{OH} are rate constants for the acid-catalyzed, the neutral, the base-promoted process. Since the concentration of water is essentially constant, its concentration can be incorporated into the neutral process constant to define a new constant

$$k_o = k_{H_2O}[H_2O] \quad (10\text{-}2)$$

The overall rate is therefore a simple first order expression when pH is fixed.

$$r_h = k_h[S] \quad (10\text{-}3)$$

where k_h is the observed hydrolysis constant

$$k_h = k_H[H^+] + k_o + k_{OH}[OH^-] \quad (10\text{-}4)$$

Other acidic and basic species besides H^+ and OH^- are potentially capable of catalyzing hydrolysis. This generalized acid or base catalysis character can be incorporated into the overall rate expression by inclusion of additional terms: first order in each specific acid and base. The result is a more complex expression for k_h

$$k_h = k_H[H^+] + k_o + k_{OH}[OH^-] + \sum_i k_{A_i}[A_i] + \sum_i k_{B_i}[B_i] \quad (10\text{-}5)$$

where k_{A_i} and k_{B_i} are the hydrolysis rate constants for acid species i and base species i, and $[A_i]$ and $[B_i]$ are the concentrations of the specific acid and base species present. Generalized acid or base catalysis of hydrolysis is discussed elsewhere [4,6,7,8].

A sorbed chemical may react quite differently than that in solution just as sorption may affect biotransformation rates.

Sorption could lead to increased, decreased, or unchanged hydrolysis rates. Sorption can be handled mathematically by writing a reaction rate expression for each phase in a manner identical to that for biotransformation. The rate of hydrolysis on the soil, r_{hs}, and in the water phase, r_{hw}, are assumed first order at a fixed pH

$$r_{hs} = k_{hs}[S_s] \tag{10-6}$$

$$r_{hw} = k_{hw}[S_w] \tag{10-7}$$

where k_{hs} and k_{hw} are the overall phase specific rate constants, and $[S_s]$ and $[S_w]$ are the concentrations of chemical on the soil and in the water, respectively.

The rate of total chemical loss, r_{hT}, can be shown to be first order in total chemical concentration $[S_T]$ if sorption is rapid.

$$r_{hT} = k_{hT}\,[S_T] \tag{10-8}$$

The overall first order "rate" constant, k_{hT}, is a function of the partition coefficient, K_p, soil to water ratio, M, and the phase specific overall hydrolysis rate constants, k_{hs} and k_{hw},

$$k_{hT} = f_1 k_{hS} + f_2 k_{hw} \tag{10-9}$$

where

$$f_1 = \frac{1}{1 + K_pM} \tag{10-10}$$

$$f_2 = \frac{K_pM}{(1 + K_pM)} \tag{10-11}$$

The presence of soils, however, may affect hydrolysis independently of sorption. Burkhard and Guth [9] observed that several herbicides are hydrolyzed in soils according to a first order relationship. They found that the rate of hydrolysis in soils was increased compared to that in pure aqueous solution, but that the rates decreased with the extent of sorption; this indicates that sorption "protected" the herbicides, and that the presence of soils accelerated hydrolysis. Konrad and Chesters [10] also showed that the hydrolysis of an organophosphate insecticide in soil was first order in chemical concentration and catalyzed in the presence of soil. In contrast to the results of Burkhard and Guth, Konrad and Chesters [10] observed that the first order rate constants were directly related to sorption (i.e., rates were proportional to the degree of sorption).

Chemical Oxidation-Reduction Reactions

The concept of chemical oxidation-reduction (redox) is more easily understood for simple inorganic reactions than for redox reactions involving organics. In the soil environment, redox reactions involving both inorganic and organic species are important.

Inorganic chemists define oxidation as the loss of electrons and increase in oxidation number while reduction is the gain in electrons and decrease in oxidation number [5]. The oxidation number of an atom represents the hypothetical charge an atom would have if the ion or molecule were to dissociate [11]. Unfortunately when dealing with organic redox reactions, these definitions are not easy to apply. While electrons are directly transferred in some redox reactions, the mechanisms of most organic reactions do not involve direct transfer but transfer of atoms [5]. Organic oxidations frequently involve a gain in oxygen and a loss in hydrogen atoms. The reverse is true for reductions.

Organic chemists have set up a series of functional groups arranged in order of increasing oxidation state to facilitate the classification of reactions as either oxidations or reductions. Table 10-1 provides a summary of the relative oxidation states of several important functional groups. Conversion of a functional group into one in a higher oxidation state category characterizes oxidation of the original group. Reduction is defined by the conversion to a group in a lower oxidation state [5].

The importance of chemical redox reactions in the soil is not well documented although laboratory model systems have shown potential importance [12,13]. Soil does contain a wide variety of substances including natural organics, metal oxides, and oxygen which may be capable of entering into redox reactions with chemicals. Several good discussions of redox reactions in soil and aquatic environments are available [14,15,16,17,18]. Further research on the significance of redox reactions in soil systems is needed particularly if transformation pathways occurring in environments

Table 10-1. Relative Oxidation States.*

		Increasing Oxidation State		
Least Oxidized				Most Oxidized
RH	ROH	RC(O)R	RCOOH	CO_2
	RCl	$(R)_2CCl_2$	$RC(O)NH_2$	CCl_4
	RNH_2	-C≡C-	$RCCl_3$	
	>C=C<*			

*Adapted from March (1977).

which do not contain significant microbial populations are to be adequately characterized.

Mechanism and Kinetics

An oxidation cannot occur without a reduction. In simple inorganic reactions, the mechanism involves the exchange of an equal number of electrons. For example, the oxidation of Fe^{+2} to Fe^{+3} by the reduction of Mn^{+4} to Mn^{+3} involves the direct exchange of a single electron

$$Fe^{+2} \rightarrow Fe^{+3} + e^{-} \qquad \text{an oxidation}$$

$$Mn^{+4} + e^{-} \rightarrow Mn^{+3} \qquad \text{a reduction}$$

$$Fe^{+2} + Mn^{+4} \rightarrow Fe^{+3} + Mn^{+3}$$

Each individual reaction is called a half-reaction (one is a reduction, the other an oxidation), and the net balanced reaction is their summation.

The concept of the additivity of half reactions can be applied even to complex organic and inorganic redox reactions using several "rules" as an aid in balancing the half reactions [11,19,20]. For example, the oxidation of glucose to carbon dioxide and water by oxygen can be written as

$$C_6H_{12}O_6 + 6H_2O \rightarrow 6CO_2 + 24H^{+} + 24e^{-} \qquad \text{an oxidation}$$

$$6O_2 + 24H^{+} + 24e^{-} \rightarrow 12\,H_2O \qquad \text{a reduction}$$

$$C_6H_{12}O_6 + 6H_2O + 6O_2 \rightarrow 6CO_2 + 6H_2O$$

In many cases, half reactions can be written that do not describe the real mechanism but are only a convenient "bookkeeping" method to correctly balance the net reaction as is the case for the oxidation of glucose cited above. Frequently, many individual elementary reactions, transient intermediates such as free radicals, and other commonly available species such as H^{+}, OH^{-}, and H_2O, are involved in the transformation of a chemical. Several different mechanisms may also lead to more than one product type.

Some of the reactions listed in the section on biotransformation show transformations of a chemical that are typical oxidation or reduction reactions (e.g., oxidative dealkylation, decarboxylation, nitro-reduction, etc.). Such reactions, of course, must be accompanied by the simultaneous reduction of some other substance if the chemical of interest is oxidized or by the oxidation of some other substance if the chemical of interest is reduced. For example, oxygen may be reduced to H_2O in an oxidation of a chemical.

Evidence for the existence of redox reactions in the soil comes from the identification of reaction products that are characteristic of redox reactions. While much is known about the mechanisms and kinetics describing the oxidation or reduction of specific organic chemicals in the laboratory by added chemical oxidants or reductants such as chlorine (Cl_2), potassium permanganate, and oxygen (O_2), very little is known about the oxidants and reductants or reaction mechanisms and kinetics of importance in soil.

No relationship has been derived from first principles to describe the general oxidation or reduction of chemicals in the soil and been validated. However, a simple second order relationship has been used to model the oxidation of a chemical in aqueous phase reactions [5,18,21,22].

$$r_{ox} = k_{ox}[ox][S] \qquad (10\text{-}12)$$

and k_{ox} is a second order rate constant under conditions where the oxidant species ox was photochemically produced and involved in the rate limiting step of the reaction mechanism. The general applicability of this or a similar rate expression describing the oxidation or reduction of a compound has yet to be shown in soil systems.

Photochemical Transformations

The term photochemical transformation is a general term referring to all reactions mediated by light through the action of a photon absorbed either by the specific chemical of interest or some naturally occurring substance. Photochemical transformations include a variety of common reactions such as oxidation, reduction, elimination, isomerization, substitution, addition, fragmentation, and hydrolysis [15,23]. Photochemical transformation of a compound may lead to one or more reaction products typical of these reactions.

Photochemical reactions have been shown to play a significant role in the transformation of several chemicals at the soil surface [14,15,24,25]. However, since light does not penetrate very far into soils, photochemical reactions do not play a significant role in the transformation of the bulk of a chemical if it is incorporated into the soil [26]. Consideration of photochemical transformations may be important in order to correctly ascertain the flux of chemical from the soil surface.

The general topic of photochemistry is discussed in the books of Turro [23] and Calvert and Pitts [27]. Several good reviews describing photolysis of environmental importance are available [2,18,28,29,30,31].

Mechanism and Kinetics

Photochemical transformations may occur by one or more processes that depend on chemical structure and the presence of other substances in the environment. The direct photolysis of a chemical occurs when the chemical that absorbs light undergoes reaction while in an excited state. Indirect photolysis occurs when one chemical called a sensitizer absorbs light energy and transfers it to another chemical which then undergoes a reaction. This type of indirect photolysis is a photosensitized process. Energy transfer can also reduce (quench) the photolysis of a target chemical if energy is transferred from the target chemical before it can react. Chemicals which deactivate the excited state of the target chemical are called quenchers. Many natural organic substances are sensitizers or quenchers.

A specific chemical may also react with an energetic intermediate produced when a natural substance absorbs light. Singlet oxygen and oxyradicals are examples of such intermediates identified as being important in aqueous systems which can react to cause the transformation of a chemical [22,31,32,33]. Their existence may also be important at the soil surface. Reactions involving these intermediates are frequently referred to as photooxygenations or photoinitiated free radical reactions (after reaction with singlet oxygen and oxyradicals, respectively). However, the mechanisms involved in photochemical transformations are frequently unclear and cannot be differentiated.

The kinetics of photochemical transformations on soils has not been extensively studied since light does not penetrate very far into soil. However, the rate of photochemical transformation, r_p, of a chemical in aqueous solution has been expressed as a first order relationship

$$r_p = K[C] \tag{10-13}$$

where C is the concentration of chemical, and K is a constant which depends on the photon flux, the light absorption coefficient, and the reaction quantum yield which expresses the efficiency of conversion of absorbed light into chemical reaction [29,34]. The first order constant is also related to the concentration of sensitizer present, if any [30]. Hautala [26] studied the photolysis of several pesticides on very thin soil samples having particles which did not block access to light and also obtained a first order rate of loss although the rates were very small in comparison to those expected in aqueous solution. Since light does not penetrate deeply into soil, first order kinetics would not be expected in the bulk of the soil.

Sorption may increase or decrease photolysis possibly by involvement of quenching agents, photosensitizing agents, or by changing the absorption spectra of a chemical [15,35,36,37]. Burkhard and Guth [25] observed a decrease in the rate of photolysis of several organophosphorus insecticides on soil with decreasing moisture content which they attributed to a protective effect of sorption. Hydrogen ion concentration may also have an effect on rates and products formed [38].

FACTORS INFLUENCING NONBIOLOGICAL TRANSFORMATIONS

A number of factors influence nonbiological reactions by either determining the availability of important reactants or modifying the reactivity. Little is known about the actual importance of many of these factors beyond conjecture based on laboratory studies involving conditions quite remote in some cases from those of environmental importance. Many of the factors listed in Table 10-2 are interrelated in complex ways. For example, the general effect on sorption needs to be considered.

Soil Moisture Content

Soil moisture commonly serves as a medium in which hydrolysis and redox reactions occur. It also affects the chemical concentration and amount sorbed. The rates of hydrolysis of some chemicals have been correlated with moisture content. Saltzman et al. [39], have shown that the rate of hydrolysis of parathion on Kaolinite, a common soil constituent, increases with moisture up to about 11 percent moisture content. At a moisture content beyond this, the rate decreased. In addition, the high solubility of oxygen in water makes soil moisture a good vehicle for transport of oxygen. However, oxidation of a compound by oxygen may be decreased if soil moisture limits oxygen availability. Surface reactions with atmospheric oxygen may still readily occur at low moisture levels.

pH

Hydrogen ion concentration may affect the rates of hydrolysis several ways:

a) Directly, because the rate of the acid-promoted process is an explicit function of hydrogen ion concentration

Table 10-2. Factors Influencing Nonbiological Reactions.

1. Soil moisture content	5. Soil clay content
2. pH	6. Metal oxides/ions and inorganic acids/bases
3. Oxygen	7. Temperature
4. Soil organic content	8. Sunlight

b) By affecting the ionization of the chemical

c) By affecting the amount of chemical sorbed

Hydrogen ion concentration is related to the hydroxide ion concentration by the ionization product of water, K_w,

$$[OH^-] = \frac{K_w}{[H^+]} \tag{10-14}$$

therefore, the generalized rate of hydrolysis (in the absence of general acidbase catalysis) may be expressed as

$$r_h = k_H[H^+] + k_o + \frac{k_{OH}k_w}{[H^+]} \% [S] \tag{10-15}$$

The exact dependence of the hydrolysis rate depends on the magnitude of each rate constant. The pH will also affect the concentration of inorganic acids and bases which may act as general catalysts.

Not all chemicals are hydrolyzed by mechanisms involving all three processes. For example, organic halide hydrolysis in aqueous solution is not significantly acid-catalyzed [1]. In addition, rate constants can differ greatly, and half-lives can vary from a few minutes to thousands of years depending on chemical type and pH. For example, Zepp et al. [40], report that the hydrolysis half-life of a butoxyethyl ester of 2,4-D decreases from one year at pH 5 to only 9 hours at pH 8. Mabey and Mill [1] report that the half-lives of the acid esters, ethyl acetate and methyl benzoate, are 20 years and 38 days, respectively, at pH 7.0.

All forms of an ionizable chemical are not hydrolyzed at the same rate. Since pH determines speciation, pH could also indirectly affect hydrolysis and give rise to rate vs. pH relationships that are complex. Ionization effects can be handled mathematically by using separate rate terms for each species and calculating the true species concentration from knowledge of the acid-base dissociation constant.

Rates of chemical redox reactions are frequently a function of hydrogen ion concentration. Some are promoted while others inhibited at higher pH values. Additionally, H^+ and OH^- may be consumed or produced during a redox reaction. Types of products formed could also be affected.

Oxygen

Oxygen (O_2) is probably the most important oxidant in soils. Atmospheric oxygen is capable of directly oxidizing many organic and inorganic compounds. Chemical oxidation by atmospheric oxygen is usually slow and has been termed autooxidation, a process that may involve trace quantities of free radicals [5]. Oxygen is not expected to affect hydrolysis reactions but could be involved in photochemical processes occurring at the soil surface (photooxygenations).

Soil Organic Matter Content

Soil organic matter sorbs many chemicals and would therefore be expected to affect the rate of hydrolysis and redox reactions. Several components of soil organic matter are known to affect hydrolysis. For example, basic amino acids catalyze the hydrolysis of organophosphate esters [15], and fulvic acids catalyze the hydrolysis of atrazine in aqueous solution [41]. Perdue [7] has pointed out that humic substances can modify hydrolysis reaction rates by general acid/base catalysis and by partitioning equilibrium that lead to a humic-bound substrate in solution with a reactivity different from that of the unbound substrate. Both catalytic and retardation effects are possible. The net effect on hydrolysis would depend on the catalytic potential, the effect of sorption, and the formation of humic-bound substrates.

The organic fraction of soil represents a potential storehouse of both oxidizing and reducing agents. Steelink and Tollin [42] have even shown that free radicals are associated with soil organic matter although their exact source and concentration remains unknown. Natural organic compounds have been found to reduce a variety of inorganic species [43]. Reactions involving xenobiotics with natural organics could also be important and should be further studied. The rate of oxidation or reduction of a specific chemical could be reduced due to competing reactions involving soil organics.

Organic matter may serve as a source of photosensitizers which effectively catalyze photochemically mediated oxidations or as a source of quenchers which inhibit such reactions. Humic material has been shown to act as a photosensitizing agent in aqueous solution [44]. Hautala [26] studied the rates of photolysis of several pesticides in soils and suggested that the slow rates observed were due to quenching of the photoexcited pesticide by soil pigments.

The presence of organic matter also may affect oxygen concentration through increased microbial activity and, therefore, indirectly may affect the rate of oxidation or reduction of a chemical.

Soil Clay Content

Soil clay may sorb chemicals and provide acidic and basic sites that catalyze hydrolysis. Because of interactions between water molecules and clay minerals, water surface films may exhibit greatly changed pH values as the water evaporates [15] and hereby may affect hydrolysis rates. The net effect of clay content would depend on the type of clay and chemical. The effect of clays on redox reactions has not been studied.

Metal Oxides/Metal Ions and Inorganic Acids/Bases

Metals are ubiquitous in the soil environment. Several alkaline earth and heavy metal ions catalyze hydrolysis. For example, Mortland and Raman [45] found that copper ions catalyzed the hydrolysis of organophosphates in clay suspensions. However, little appears to be known about the importance of metal catalysis in the environment. The concentrations of metal ions normally found in water may be too low in concentration to significantly affect hydrolysis, or they are complexed with organic substances in such a manner that they are not effective catalysts [1]. Soil waters may also contain a wide variety of weak inorganic acids and bases including carbonates, silicates, phosphates, sulfides, and amines that could potentially affect hydrolysis by acting as general acid or base catalyst [8].

Metal ions can act as reactants in redox reactions. Fleck and Haller [46] demonstrated the conversion of DDT to DDE in the presence of $FeCl_3$ and $AlCl_3$. Glass [47] investigated the reduction of DDT to DDD in soils and proposed that the reaction involved transfer of electrons from reduced organic matter to DDT via a Fe^{+2} - Fe^{+3} redox couple in the absence of oxygen. Stone and Morgan [12, 13] have shown that many organic compounds may be oxidized by manganese (III) and manganese (IV) oxides. Ferric salts have also been shown to act as photosensitizing agents [14,48].

Temperature

The rates of all elementary chemical reactions increase as temperature increases. However, since sorption may play a role in the overall rate of reactions, the effect of temperature on sorption should also be considered.

Elementary rate constants are frequently correlated with temperature using the Arrhenius expression

$$k = A e^{-E_A/RT} \tag{10-16}$$

where k is the rate constant, E_A is the activation energy, R is the gas constant, T is absolute temperature, and A is a pre-exponential factor. The Arrhenius expression should be applied only to rate constants for <u>individual</u> processes and not necessarily applied to the overall observed rate constant. For example, only the individual acid, neutral, and base hydrolysis constants (k_H, k_O, and k_{OH}) should be correlated with temperature using the Arrhenius equation and not by using the overall observed hydrolysis constant (K_h or K_{hT}).

The Arrhenius expression can be linearized by taking the logarithm of both sides of equation (10-16) to yield

$$\log k = \log A - E_A/RT \tag{10-17}$$

which provides a useful way to present temperature correlations graphically by plotting log k vs 1/T. Mabey and Mill [1] suggested the correlation of aqueous hydrolysis rate data using a form which more accurately describes the effect of temperature

$$\log k = \frac{-A}{T} + B \log T + C \tag{10-18}$$

where A, B, C are constants.

Increased temperature generally increases the rate of redox reactions. However, since many redox reactions are not elementary and may involve competing reactions that consume oxidants, the exact effect cannot be stated *a priori*. Free radical oxidations are not strongly temperature dependent. Temperature may also affect the concentration of oxygen dissolved in soil water and therefore may affect reactions involving oxygen. The solubility of oxygen increases with decreasing temperature.

Sunlight

Several characteristics of sunlight influence photochemical reactions [15,34]. The energy content of light depends on its wavelength, and only photons having sufficient energy can lead to a reaction. Additionally, not all wavelengths of light are absorbed by chemicals. The rate of reaction may be a function of the rate at which photons impinge on a surface (intensity). Both sunlight wavelength and intensity are a function of time of day, season, and geographical location.

ESTIMATION OF RATE PARAMETERS

Field vs. Laboratory Studies

Evaluation of the importance of nonbiological transformations under field conditions is difficult because of concurrent biological processes leading to chemical transformation and in some cases to products that may be identical to those produced by a chemical reaction. Laboratory tests using sterile soil samples are required if the importance of abiotic transformation pathways is to be ascertained and if rate constants are to be estimated.

A minimum objective should be to determine if significant nonbiological transformations are possible at all. For example, hydrolysis can be studied in water with or without soil present. The pH can easily be modified by acid or base addition. The relative stability of a chemical in the presence and absence of oxygen will indicate the probable importance of oxygen in the field environment. Oxygen concentration can be controlled by appropriate mixing of nitrogen and oxygen (and CO_2 if needed). Redox potentials can be measured as an indicator of the relative oxidation state of a soil [49]. Light/dark experiments can be conducted to ascertain if photochemical transformations can occur. Soil to which a chemical has been applied can be exposed to natural or artificial light. General laboratory methods applicable to the study of nonbiological transformations have been discussed by Howard et al. [50]. Laboratory methods for the study of hydrolysis have been discussed by Mill et al. [18], and Wolfe et al. [3,51]. Several approaches to studying photochemical reactions on soil have been used by various researchers [16,17,25,26,50,52,53].

Measurement at Test Chemical Disappearance

Ideally, not only should the concentration of test chemical be measured, but also the identity and concentration of the products should be measured. The rate of product formation should equal the rate of chemical transformation. If hydrolysis is suspected, then the products formed should be characteristic of this. The same is true for redox reactions. The possibility of multiple products also complicates interpretation of the results.

Measured transformation rates in the soil are based on total chemical concentration, usually determined by solvent extraction. However, experimental methodologies can be developed (discussed in the next section) which could be used to determine the relative importance of the soil fraction in promoting or inhibiting a reaction.

Nonbiological Transformation Rate Expressions and Constants

Goring et al. [54], and Hamaker [55,56] have discussed general empirical rate expressions (a power rate law and a hyperbolic expression) frequently used to describe various degradative processes occurring in the soil (see discussion of biotransformation kinetics in Chapter 9). However, nonbiological transformation rate expressions in soil systems have been expressed primarily as a first order rate relationship

$$r = kC \tag{10-19}$$

where k is a first order rate constant, and C is the chemical concentration. In a batch system, equation (10-19) may be integrated to yield an equation relating concentration kC at time t to the initial concentration C_o.

$$\ell n\ C/C_o = -kt \tag{10-20}$$

First order rate constants can be obtained from batch experiments by plotting this integrated expression or from measurements of the test chemical half-life, $t_{1/2}$

$$k = .693/t_{1/2} \qquad (10\text{-}21)$$

Ou et al. [57], and Rao and Davidson [58] have provided an extensive summary of first order rate constants and half-lives for chemicals in the soil environment. However, much of the data was obtained in the presence of microorganisms and therefore should not be interpreted to include only chemical transformations.

The rate of hydrolysis is expected to follow a first order relationship in soils as previously discussed. Several hydrolysis "rate" constants can be determined in soil systems. The utility of each would depend on the formulation of the soil model being used. The simplest and most site specific constant is the overall hydrolysis rate constant, k_{hT}, characterizing the first order loss in total chemical concentration.

The pure aqueous phase constants k_h, k_H, k_o, and k_{OH} can be obtained from batch laboratory experiments conducted in the absence of soil using buffered, sterile water. However, Perdue and Wolfe [8] have pointed out that buffers may accelerate the rate of hydrolysis. They have provided a model which can be used as a guide in selecting appropriate buffers. The value of the first order constant k_h can be obtained in a manner similar to that of k_{hT}. Using three pH values, k_H, k_o, and k_{OH} can be obtained from the simultaneous solution of three linear equations each corresponding to equation (10-4). Constants characterizing the effect of humic substances and inorganics can also be evaluated [7]. Mabey and Mill [1] have summarized rate constants for a number of organic compounds in water. Perdue and Wolfe [8] have presented a model which can be used to estimate the potential catalytic effect of inorganic acids and bases. Perdue [7] has presented a model allowing the estimation of the potential effect of humic material on the rate of hydrolysis.

The phase specific overall hydrolysis constants k_{hw} and k_{hs} can (at least in principle) be obtained from experiments at a fixed pH utilizing different soil-to-water ratios (M values) if the partition coefficient (K_p) is known. Two simultaneous linear equations derived from equation (10-9) using measured k_{hT} values and calculated f_1 and f_2 values can be solved to yield estimates of k_{hs} and k_{hw}. The pH dependency could be further investigated as previously discussed.

Chemical redox reactions have been studied primarily in model systems in the laboratory where oxygen concentration and other parameters can be controlled and evaluated. Most laboratory studies utilize conditions greatly different from those in the environment. Observations are generally made regarding persistence.

First order rate constants can be extracted from the data even if the data indicates that a more complex relationship exists. First order constants are expected to be particularly site specific since the concentration of oxidant or reductant is not explicitly accounted for. A second order relationship could possibly be developed if the oxidant/reductant in the soil can be identified and their concentration measured. However, at present this is beyond the sophistication of any available soil model.

Photochemical transformation studies which encompass redox reactions have been conducted under field and laboratory conditions. Field studies have focused on the determination of sunlight effects on relative persistence. No effort is usually made to quantify rates in terms of any model. Laboratory studies have likewise not been aimed at modeling but have been aimed primarily at determination of relative persistence and product formation.

LIMITATIONS

Many of the limitations encountered in the measurement and use of biotransformation constants also apply to those characterizing nonbiological reactions.

Multiple Processes and Pathways

Physical phenomena such as volatilization, other chemical processes, and biotransformation can simultaneously occur causing an overestimation of rate constants. The use of soil sterilization, while eliminating biotic factors, may affect some property that affects a chemical reaction. For example, Konrad and Chesters [10] observed a decrease in the rate of hydrolysis of an organophosphate insecticide after electron-beam irradiation was used to sterilize the soil. This decrease was attributed to changes in the adsorption of the insecticide, not from the retardation of microbial processes.

The inability to resolve nonbiological transformations into their component reactions would lead to the determination of rate "constants" which are especially empirical. Additionally, several independent reaction pathways each producing different products could exist giving rise to a complex rate dependence.

Bound Chemical Residues

Inability to measure all of the applied chemical could result in a calculated rate constant too high if it is based on loss of a parent compound that forms a bound residual or could result in a low estimate of a rate constant if it is based on formation of a known product that forms a bound residue.

Specificity of Constants

As nearly as possible, constants characterizing transformation rates should be used under those soil conditions and concentration ranges at which they were obtained since so many unknown factors can affect chemical reactions. Unfortunately, laboratory studies frequently are conducted under conditions very different from those encountered in the field.

REFERENCES

1. Mabey, W., and T. Mill. "Critical Review of Hydrolysis of Organic Compounds in Water Under Environmental Condition," J. Phys. Chem. Ref. Data 7:383-415 (1978).

2. Harris, J. C. "Rate of Hydrolysis," in Handbook of Chemical Property Estimation Methods: Environmental Behavior of Organic Compounds, W. J. Lyman, W. F. Reehl, and D. H. Rosenblatt, Eds. (New York: McGraw-Hill Book Co., 1982).

3. Wolfe, N. L., R. G. Zepp, J. A. Gordon, G. L. Baughman, and D. M. Cline. "Kinetics of Chemical Degradation of Malathion in Water," Environ. Sci. Technol. 11:88-93 (1977).

4. Lowry, T. H., and K. S. Richardson. Mechanism and Theory in Organic Chemistry (New York: Harper and Row, 1976).

5. March, J. Advanced Organic Chemistry, 2nd ed. (New York: McGraw-Hill Book, Co., 1977).

6. Bell, R. P. Acid-Base Catalysis (Oxford: Clarendon Press, 1941).

7. Perdue, E. M. "Association of Organic Pollutants with Humic Substances: Partitioning Equilibria and Hydrolysis Effects in Aquatic and Terrestrial Humic Materials," R. F. Christman and E. G. Gjessing, Eds. (Ann Arbor, MI: Ann Arbor Science Publications, 1983).

8. Perdue, E. M. and N. L. Wolfe. "Prediction of Buffer Catalysis in Field and Laboratory Studies of Pollutant Hydrolysis Reactions," Environ. Sci. Technol. 17:635-642 (1983).

9. Burkhard, N., and J. A. Guth. "Chemical Hydrolysis of 2-chloro-4,6-bis(alkylamino)-1,3,5-triazine Herbicides and Their Breakdown in Soil Under the Influence of Adsorption," Pestic. Sci. 12:45-52 (1981).

10. Konrad, J. G., and G. Chesters. "Degradation in Soils of Ciodrin, an Organophosphate Insecticide," J. Agr. Food Chem. 17:226-230 (1969).

11. Stumm, W., and J. J. Morgan. Aquatic Chemistry, 2nd ed. (New York: John Wiley & Sons, 1981).

12. Stone, A. T., and J. J. Morgan. "Reduction and Dissolution of Manganese (III) and Manganese (IV) Oxides by Organics. 1. Reactions with Hydro-Quione," Environ. Sci. Technol. 18:450-456 (1984).

13. Stone, A. T., and J. J. Morgan. 1984b. "Reduction and Dissolution of Manganese (III) and Manganese (IV) Oxides by Organics. 2. Survey of the Reactivity of Organics," Environ. Sci. Technol. 18:617-624 (1984).

14. Helling, C. J., P. C. Kearney, and M. Alexander. "Behavior of Pesticides in Soil," Adv. Agron. 23:147-240 (1971).

15. Crosby, D. G. "Nonbiological Degradation of Herbicides in the Soil," in Herbicides: Physiology, Biochemistry, Ecology, L. J. Audus, Ed. (New York: Academic Press, 1976).

16. Plimmer, J. R. "Photolysis of TCDD and Trifluralin on Silica and Soil," Bull. Environ. Contam. Toxic. 20:87-92, (1978).

17. Plimmer, J. R. "Degradation Methodology: Chemical-Physical Effects," EPA-600/9-79-012, Proceedings of the Workshop on Microbial Degradation of Pollutants in Marine Environments, Washington, DC (1978).

18. Mill, T., W. R. Mabey, D. C. Bomberger, T. W. Chou, D. C. Hendry, and J. H. Smith. "Laboratory Protocols for Evaluating the Fate of Organic Chemicals in Air and Water," EPA-600/3-82-022 (Washington, DC: U.S. EPA, 1982).

19. Sawyer, C. N., and P. L. McCarty. Chemistry for Environmental Engineers, 3rd ed. (New York: McGraw-Hill Book Co., 1978).

20. Snoeyink, V. L., and P. Jenkins. "Water Chemistry, (New York: John Wiley & Sons, 1980).

21. Smith, J. H., W. R. Mabey, N. Bohonos, B. R. Holt, S. S. Lee, T. W. Chou, D. C. Bomberger, and T. Mill. "Environmental Pathways of Selected Chemicals in Freshwater Systems. Part I: Background and Experimental Procedures," EPA-600/7-77-113 (Washington, DC: U.S. EPA, 1977).

22. Mill, T., D. G. Hendry, and H. Richardson. "Free-Radical Oxidants in Natural Waters," Science 207:886-887 (1980).

23. Turro, N. J. Molecular Photochemistry (Amsterdam, NY: W. A. Benjamin, 1965).

24. Lichtenstein, E. P., K. R. Shulz, T. W. Fuhremann, and T. T. Liang. "Degradation of Aldrin and Heptachlor in Field Soils During a Ten-Year Period," J. Agr. Food Chem. 18:100-106 (1970).

25. Burkhard, N., and J. A. Guth. "Photolysis of Organophosphorus Insecticides on Soil Surfaces," Pest. Sci. 10:313-319 (1979).

26. Hautala, R. R. "Surfactant Effects on Pesticide Photochemistry in Water and Soil," EPA-600/3-78-060 (U.S. EPA, 1978).

27. Calvert, J. G. and J. N. Pitts. Photochemistry (New York: John Wiley & Sons, Inc., 1966).

28. Zifiriou, O. C. "Marine Organic Photochemistry Previewed," Marine Chemistry 5:497-522 (1977).

29. Baughman, G. L., and R. R. Lassiter. "Prediction of Environmental Pollutant Concentration," Special Technical Publication 657 (Philadelphia, PA: ASTM, 1978).

30. Zepp, R. G., and G. L. Baughman. "Prediction of Photochemical Transformation of Pollutants in the Aquatic Environment," in Aquatic Pollutants: Transformations and Biological Effects, O. Hutzinger, I. H. Van Lelyveld, and B. C. J. Zoetman, Eds. (New York: Pergamon Press, 1978).

31. Miller, S. "Photochemistry and Natural Water Systems," Environ. Sci. Technol. 17:568A-570A (1983).

32. Zepp, R. G., N. L. Wolfe, G. L. Baughman, and R. C. Hollis. "Singlet Oxygen in Natural Waters," Nature 267:421-423 (1977).

33. Wolfe, C. J. M., M. T. H. Halmans, and H. V. Vander Heijde. "The Formation of Singlet Oxygen in Surface Waters," Chemosphere 10:59-62 (1981).

34. Zepp, R. G., and D. M. Cline. "Rates of Direct Photolysis in Aquatic Environments," Environ. Sci. Technol 11:359-366 (1977).

35. Ivie, G. W., and J. E. Casida. "Sensitized Photodecomposition and Photosensitizer Activity of Pesticide Chemicals Exposed to Sunlight on Silica Gel Chromatoplates," J. Agr. Food Chem. 19:405-409 (1971).

36. Rosen, J. D. "The Photochemistry of Several Pesticides," in: Environmental Toxicology of Pesticides, F. Matsumura, Ed. (New York: Academic Press, 1972).

37. Miller, G. C., and R. G. Zepp. "Photoreactivity of Aquatic Pollutants Sorbed on Suspended Sediments," Environ. Sci. Technol. 13:860-863 (1979).

38. Crosby, D. G., and Leitis. "The Photodecomposition of Triafluralin in Water," Bull. Environ. Contam. Toxic. 10:237-241 (1973).

39. Saltzman, S., U. Mingelgrin, and B. Yaron. "Role of Water in the Hydrolysis of Parathion and Methylparathion on Kaolinite," J. Agric. Food Chem. 24:739-743 (1976).

40. Zepp, R. G., N. L. Wolfe, G. L. Baughman, and J. A. Gordon. "Dynamics of 2-4,D Esters in the Aquatic Environment: Hydrolysis and Photodegradation, Presented at the 167th American Chemical Society Meeting, Los Angeles, CA, 1974.

41. Khan, S. U. "Kinetics of Hydrolysis of Atrazine in Aqueous Fulvic Acid Solution," Pest. Sci. 9:39-43 (1978).

42. Steelink, C., and G. Tollin. "Free Radicals in Soil," in Soil Biochemistry, A. D. McLarenand and G. H. Peterson, Eds. (New York: Marcel-Dekker, Inc., 1967).

43. Skogerboe, R., and S. A. Wilson. "Reduction of Ionic Species by Fuluric Acid," Anal. Chem. 53:228-232 (1981).

44. Zepp, R. G., G. L. Baughman, and P. F. Schlotzhaver. "Comparison of Photochemical Behavior of Various Humic Substances in Water. II. Photosensitized Oxygenations," Chemosphere 10:119-126 (1981).

45. Mortland, M. M., and K. V. Raman. "Catalytic Hydrolysis of Some Organic Phosphate Pesticides by Copper (II)," J. Agr. Food Chem. 15:163-167 (1967).

46. Fleck, E. E., and H. L. Haller. "Compatibility of DDT with Insecticides, Fungicides and Fertilizers," Ind. Eng. Chem. 37:403-405 (1945).

47. Glass, B. L. "Relation Between Degradation of DDT and the Iron Redox System in Soils," J. Agr. Food Chem. 20:324-327 (1972).

48. Koster, R., and K. D. Asmur. "Reactions of Fluoridated Benzenes with Hydrated Electrons and Hydroxyl Radicals in Aqueous Solution," J. Phys. Chem. 77:749-755, (1973).

49. Bohn, H. L. "Redox Potentials," Soil Sci. 112:39-45 (1971).

50. Howard, P. H., J. Saxena, P. R. Durkin, and L. T. Ou. "Review and Evaluation of Available Techniques for Determining Persistence and Routes of Degradation of Chemical Substances in the Environment," EPA-560/5-75-006 (U.S. EPA, 1975).

51. Wolfe, N. L., W. C. Steen, and L. A. Burns. "Phthalate Ester Hydrolysis: Linear Free Energy Relationships," Chemosphere 9:403-408 (1980).

52. Parochetti, J. V., and G. W. Dec., Jr. "Photodecomposition of Eleven Dinitroaniline Herbicides," Weed Sci. 26:153-156 (1978).

53. Smith, C. A., Y. Iwata, and F. A. Gunther. "Conversion and Disappearance of Methidathion on Thin Layers of Dry Soil," J. Agric. Fd. Chem. 26:959-962 (1978).

54. Goring, C. A. I., D. A. Laskowski, J. Hamaker, and R. W. Miekle. "Principles of Pesticide Degradation in Soil," in Environmental Dynamics of Pesticides, R. Hague and V. H. Freed, Eds. (Plenum Press, 1975).

55. Hamaker, J. W. "Mathematical Predictions of Cumulative Levels of Pesticides in Soil," Adv. Chem. Series. 60:122-131 (1966).

56. Hamaker, J. W. "Decomposition: Quantitative Aspects," in Organic Chemicals in the Soil Environment, C. A. I. Goring and J. W. Hamaker, Eds. (New York: Marcel-Dekker, Inc., 1972).

57. Ou, L. T., J. M. Davidson, P. S. C. Rao, and W. B. Wheeler. "Pesticide Transformations in Soils," in Retention and Transformation of Selected Pesticides and Phosphorous in Soil-Water Systems: A Critical Review, P. S. C. Rao and J. M. Davidson, Eds, EPA-600/3-82-060 (U.S. EPA, 1982).

58. Rao, P. S. C., and J. M. Davidson. "Estimation of Pesticide Retention and Transformation Parameters Required in Non-Point Source Pollution Models," in Environmental Impact of Nonpoint Source Pollution, M. R. Overcash and J. M. Davidson, Eds. (Ann Arbor, MI: Ann Arbor Science Publishers, 1980).

CHAPTER 11

SPATIAL VARIABILITY OF SOIL PROPERTIES

W. A. Jury
University of California-Riverside, Riverside, California

INTRODUCTION

Soils vary significantly from point to point in their structural properties, textural composition, and mineralogical constituents. As a result, virtually all of the parameters characterizing the transport processes discussed in this report vary both laterally and vertically in an undisturbed soil profile of field size. Consequently, in the field setting, the task of modeling becomes considerably more complex than for laboratory scale processes. However, even though the transport processes to be characterized will inevitably display lateral and vertical variation, it is not reasonable to expect that three-dimensional models capable in principle of describing this point-to-point variability could be calibrated by any conceivable combination of measurements. As a result, field-scale models, at least for processes in which large surface areas are treated relatively uniformly (i.e., agricultural operations, large waste holding ponds), will have to be one-dimensional. Thus, replicated measurements of representative parameters for these one-dimensional models where large field areas are involved must be substantial enough so that their mean values give a representative average over the field. Furthermore, the uncertainty of the measurement, represented by the sample variance or coefficient of variation, should be recorded as well as the mean value to indicate the precision of the mean estimate and to act as an indication of the extent to which the assumption of one dimensionality is valid for the field in question.

A consequence of the transition from laboratory simulation and experimentation to variable natural environments is that the scale of the averaging process becomes very important. Simulation of processes occurring in soil tank lysimeters, for example, will be more difficult and less exact than representation of transport through laboratory columns but will be considerably less difficult than representation of the average downward movement from a large field.

There are a number of implications of field variability and the admitted futility of constructing a three-dimensional chemical transport model. Most significantly, one must abandon any hope of attempting to estimate a continuous spatial pattern of chemical emissions at each point in space and must rely instead on prediction of characteristic flow indices (i.e., mean solute concentration at depth z) which represent averages across the entire cross-sectional area of the field. A second significant implication is the possibility for extreme deviations from average movement to occur over a relatively small fraction of the total cross-sectional area. Thus, there may be plumes of solute moving downward through cracks, crevices, or other natural structural voids which could carry potentially harmful pollutants to great depths and yet would be difficult to detect by a point measurement of concentration.

Dilemmas posed by the extent of observed variability have caused researchers recently to propose that transport be modeled by stochastic as opposed to deterministic hypotheses. In a stochastic model, prediction of concentrations at depths as a function of time are interpreted as an index of the relative probability of finding chemical at the given depth [1]. This stochastic philosophy may be better suited to the task of assessing extreme movement of hazardous chemicals than the deterministic models would be.

Characterization of the spatial variability of a parameter is not completely straightforward. The traditional approach is to assume that all replicated measurements of the property are statistically independent and to represent the property variation by specifying its sample mean and sample variance or coefficient of variation. However, recent investigations have shown that lateral correlations exist between transport properties measured near to each other [2,3,4], and that this statistical dependence should be taken into account when calculating sampling strategy [5] or when interpolating between measurements [6].

Two distinct philosophies are currently in evidence in the ongoing research dealing with spatial variability. The first approach, called geometric scaling theory [7,8,9], uses certain standardized variables to scale the differential equations describing transport and relates the standardized variables to some measurable or definable property of each local site of a heterogeneous field. When these variables have been identified, the task of characterizing variability is reduced to calculating the statistical and spatial distribution of these scaling parameters. The only current application of scaling theory used so far is the so-called geometric similitude scaling [10] which assumes magnified scale heterogeneity in which each individual soil location is regarded as a magnified version of a reference location which may be characterized by a single microscopic length parameter. This length parameter has a predictable influence on transport properties such as hydraulic conductivity and matric potential. Thus, the statistical distribution of the scaling factors combined with a single set of reference transport coefficient functions allows one to make repeated calculations of flow properties from a single computation at the reference site and also to simulate transport processes for all locations at once by solving the scaled transport equations [11].

The second hypothesis for treating spatial variability is to regard the various parameters relevant to a field-wide description of transport as random variables characterized by a mean value and a randomly fluctuating stochastic component. A sampling at a point thus reveals a single momentary snapshot of these fluctuating properties which may be analyzed by statistical techniques designed to detect spatial correlations in order to yield information about the spatial distribution of the statistical fluctuations. The stochastic solute transport models which treat solute concentration as a random variable then use the global statistical properties calculated from this analysis to predict large-scale macrodispersion and macrovelocity parameters which are used in the asymptotic (long time) description of solute transport [12,13].

The discussion to follow will begin with a description of the known experimental information characterizing the extent of variability of key soil water and solute parameters. Next, the experimental evidence in support of the two key statistical models, scaling theory and regionalized variable analysis, will be discussed and summarized. Finally, the implications of spatial variability on parameter measurement will be analyzed using both the assumption of statistical independence and the assumption of statistical dependence. Much of the tabular material in this chapter is adapted from [14].

SUMMARY OF OBSERVED VARIABILITY

Study 1 - Soil Matrix and Water Retention Properties

Many of the parameters needed for transport models have properties which are dominated by the bulk characteristics of the solid matrix of the soil. As such, their lateral variability is relatively small, reflecting the uniformity of the soil formation processes.

Table 11-1 summarizes the available information from field studies of these so-called matrix properties which is summarized by giving the field mean and sample coefficient of variation for various parameters along with the soil texture and field size where the measurements were made. Also included is the method of measurement used in the study, the number of replicate measurements used in the averaging process, and the literature citation.

For the most part, properties in this group show low to moderate variation among replicates irrespective of field size or soil

Table 11-1. Field Studies of Soil Matrix and Water Retention Properties.

Parameter	Mean	CV (%)	Field Size (ha)	Soil Texture	Number of Repli-cates	Method of Measurement	Reference
Porosity	0.45	11	150	clay loam	120	Water content at zero suction	[15]
Porosity	0.37	11	0.8	sand	120	Not given	[16]
Porosity	0.53	7	.03	clay loam	20	Water content at zero suction	[17]
Porosity	0.42	10	0.4	loamy sand	12	Water content at zero suction	[18]
Bulk Density	1.36	7	150	clay loam	120	Undisturbed cores	[15]
Bulk density	1.30	7	15	sandy loam	64	Not given	[19]
Bulk density	1.20	26	3.8	sandy loam	30	Measure volume of plastic-lined hole	[20]
Bulk density	1.47	9	1.3	sandy loam	192	Undisturbed cores	[21]
Bulk density	1.26	6	0.5	silty clay	144	Undisturbed cores	[21]
Bulk density	1.47	6	0.5	silt loam	72	Undisturbed cores	[21]
Bulk density	1.65	3	0.34	sand	5	Undisturbed cores	[22]
Bulk density	1.59	6	91.6	sand	5	Undisturbed cores	[22]
Bulk density	1.20	15	40	clay loam	36	Undisturbed cores	[23]
% sand/% clay	24/45	15/33	150	clay	480	Hydrometer	[15]
% sand/% clay	17/32	32/16	85	silty clay loam	100	Light scattering	[24]
% sand/% clay	59/12	37/53	15	sandy loam	64	Not given	[19]
% sand/% clay	83/9	3/34	91.4	loamy sand	5	Not given	[22]
% sand/% clay	65/28	8/18	0.28	sandy clay loam	35	Not given	[25]

(continued)

Table 11-1. (continued).

Parameter	Mean	CV (%)	Field Size (ha)	Soil Texture	Number of Repli- cates	Method of Measurement	Reference
.1 bar water content	.37	4	85	clay loam	100	Hanging water table	[24]
.09 bar water content	.37	17.6	150	clay loam	120		[15]
.1 bar water content (θg)	.27	20	15	sandy loam	64		[19]
.1 bar water content	.45	15	40	clay loam	36		[23]
15 bar water content	.166	14.4	85	clay loam	900	Pressure plate	[24]
15 bar water content	.041	45	1.3	sandy loam	172	Pressure plate	[21]
15 bar water content	.193	14	0.5	silty clay	144	Pressure plate	[21]
15 bar water content	.074	19	3.3	silt laom	192	Pressure plate	[21]
15 bar water content (θg)	.095	33	15	sandy loam	64	Pressure plate	[19]
pH	6.1	15	.04	clay loam	1,040	Pressure plate	[26]
pH	6.4	7	.02	loam	640	Pressure plate	[26]
pH	5.8	9	.04	sandy loam	208	Pressure plate	[26]
pH	8.2	2	85	clay loam	100	Pressure plate	[24]
K_D (cm^3/g)	2.01	31	0.64	loamy sand	36	Batch equlibrium	[27]

type. Also, there are no obvious relationships between parameter value and soil texture or between parameter variability and soil texture. The limited evidence presented here suggests that one could associate an average coefficient of variation (e.g., 10 percent for porosity) with a given parameter irrespective of the particular field properties.

Study 2 - Water Transport Parameters

Table 11-2 summarizes observed variability in field experiments for water transport parameters, including saturated hydraulic conductivity, infiltration rate, and hydraulic conductivity-water content or hydraulic conductivity-matric potential functions. In contrast to the static soil properties covered in Table 11-1, this group of properties is characterized by variability of the order of 100 percent or greater in many cases and by significant differences in the extent of variability of a given property between different investigations. For example, saturated hydraulic conductivity coefficient of variation ranged from a low of 48 percent to a high of 320 percent for different Soil Conservation Service investigations in the Imperial Valley of California [28].

Infiltration rate in most of the studies varied less than saturated hydraulic conductivity, and this is owing partly to the fact that most investigations measured infiltration rate over the early stage of infiltration when matric potential gradients as well as gravity were assisting in the flow. This may have acted as a homogenizing influence. The final group of properties in Table 11-2, the parameters for the unsaturated hydraulic conductivity-water content functions and hydraulic conductivity-matric potential functions, showed enormous variability in the investigations summarized. The large variation in these functional parameters should act as a warning against using a single function to represent a large field area. As a general rule, the individual transport parameters measured at a particular location apply only to a small region around that location. Thus, predictions of models using measured values of the parameters also apply only to the vicinity of the calibration.

Study 3 - Solute Concentration Variations

Variability of solute concentration among replicates is critically important for the problem of calibrating and validating solute transport models. Table 11-3 shows summaries of the extent of concentration variation among replicates observed in 10 field experiments involving either analysis by soil solution samplers or by soil coring. The first four studies refer to solute transport experiments in which pulses of chemical were applied to the surface. The coefficient of variation reported refers to variation around the peak height maximum which contains the most critical information for model validation. These coefficients of variation lie between approximately 60 and 130 percent; in the absence of

Table 11-2. Field Studies of Water Transport Properties.

Parameter ($cm\ d^{-1}$)	Mean	CV (%)	Soil Texture	Field Size (ha)	Number of Measurements	Method of Measurement	Reference
Saturated K	20.6	120	Clay loam	150	120	Steady infiltration 20 plots x 6 depths	[15]
Saturated K	168	190	Sandy loam	15	64	Lab permeameter	[19]
Saturated K	316	69	Sand	0.8	90	In Situ air-entry permeameter	[16]
Saturated K	84	69	Loamy sand	0.4	12	Lab permeameter	[18]
Saturated K	3.6	48	Silty clay loam	Composite SCS date	33	Lab permeameter	[28]
Saturated K	18.9	103	Coarse	for a given soil series	330	Lab permeameter	[28]
Saturated K	11.0	118	Fine	in Imperial Valley, CA	287	Lab permeameter	[28]
Saturated K	6.9	92	Silty clay		339	Lab permeameter	[28]
Saturated K	28.1	320	Very coarse		36	Lab permeameter	[28]
Saturated K	55.6	118	Coarse		352	Lab permeameter	[28]
Saturated K	71.2	105	Loamy sand		121	Lab permeameter	[28]
Saturated K	98.5	81	Loamy sand	91.6	5	Lab permeameter	[22]
Saturated K	24.1	178	Sandy laom	91.6	5	Lab permeameter	[22]
Saturated K	203	50	Silt loam	9.6	26	0-30 cm infiltration (double ring)	[29]
Infiltration	14.6	94	Clay loam	150	20	Steady state	[15]

(continued)

Table 11-2. (continued).

Parameter ($cm\ d^{-1}$)	Mean	CV (%)	Soil Texture	Field Size (ha)	Number of Measurements	Method of Measurement	Reference
Infiltration rate	16.8	40	Loam	0.9	1,280	Steady infiltration (double ring)	[30]
Infiltration rate	6.6	71	Silty clay loam	.004	625	Adjacent infiltrometers along transect	[31]
Infiltration rate	8.5	56	Silty clay	.004	125		
Infiltration rate	8.5	23	Silty clay loam	.004	25		
Infiltration rate	47	79	7 series	100 ha	20	Double ring infiltrometer	[32]
Infiltration rate	263	97	7 series	100 ha	15	Inverse auger hole method 150 cm	[32]
Unsaturated $K(\theta)$ $K(\theta)=K_o\exp(\beta[\theta-\theta_o])$							
K_o	22.5	343	Clay loam	150	20	Instantaneous profile method	[33]
β	14.6	64					
K_o	4.6	235	Clay loam		20	Unit gradient method	[33]
β	89.1	41					
K_o	9.6	76	Silt laom		611	Unit gradient method	[34]
β	65.4	37					
K_o	4.0	46	Loam	.66	24	Instantaneous profile method 4 plots x 6 depths	[35]
β	32.9	19					

Table 11-3. Field Studies of Chemical Concentrations.

Chemical	Origin of Chemical	Measurement Depth (m)	Soil Texture	Field Size (ha)	Number of Replicates	CV	Method of Measurement	Reference
1. Chloride	3 cm pulse	0.65 1.15	Loamy sand	0.64		127 79 76	Soil cores	[36]
2. Bromide	1 cm pulse	0.3 0.6 0.9	Loamy sand	0.64		118 90 89	Solution samplers	[36]
3. Bromide	0.5 cm pulse	0.6	Sandy loam	0.2		100	Soil cores	[37]
4. Chloride	Surface application	0.12 0.32	Loamy sand	0.3		67 61	Soil cores	[38]
5. EC(1:1 extract)	Native	0-.025 .075-.15	Mankos shale			128 263	Soil samples	[39]
6. EC(sat extract)	Sugarcane Plantation	0-1 0-1 0-1	Clay loam-loam Clay loam-loam Clay loam-loam		150 440 445	91 75 225	Soil cores	[40]

(continued)

Table 11-3. (continued).

Chemical	Origin of Chemical	Measurement Depth (m)	Soil Texture	Field Size (ha)	Number of Replicates	CV	Method of Measurement	Reference
7. Chloride	Fertilized fields	1.8-6	Sandy loam-loam		16	22	Soil cores consolidated to depth average below root zone	[41]
		1.8-6	Sandy loam-loam		8	30		
		1.0-3.6	Loam-sandy loam		20	101		
		1.0-4.2	Loam-sandy loam		13	21		
8. Chloride	Native	Surface to bedrock ~20 cm	Various		100-200	12-70	Summary of 9 catchment areas Average concentration in core	[42]
9. Chloride	Manure fertilized fields	1.5-6.3	Sandy loam		81	19	Average concentration in entire vertical profile	[43]
		1.5-6.3	Sandy loam		14	19		
		1.5-6.3	Clay loam		14	66		
10. Chloride	Irrigated-fertilized fields	.6-1.2	Loam		--	60	Soil samples	[44]
		.9-1.5	Fine sandy loam		8	102	Soil samples	
		.6-1.2	Fine sandy loam		2	87	Soil samples	

more comprehensive information these coefficients could be used as an index to predict anticipated variation in future studies as a means of designing monitoring networks. The other studies in the table relate more to characterization of salinity and involve either assessment of native conditions (Studies 5, 6, 8) or averaging over entire profiles for assessment of nutrient loading below irrigated agricultural fields (studies 7, 9, 10). Because they represent native conditions or are the result of depth averaging their coefficient of variation is not strictly comparable to that of the transport experiments (studies 1-4).

Study 4 - Solute Velocity Variations

Since mass flow is the principal mechanism governing solute movement, the key parameter to measure or assess under field conditions is the apparent solute velocity. Not surprisingly, this parameter has been observed to vary substantially when large area experiments are conducted. Table 11-4 summarizes six studies in which the statistical distribution of this property or of related properties was obtained. The table summarizes either solute velocity, net applied water required to reach a given depth, or the depth of peak height solute concentration in a set of replicated soil cores. Although these parameters are not strictly comparable, it may be shown that when log transformed and fitted to a lognormal distribution, their log variance σ^2 should be very similar for a given study [45]. Thus, values of σ^2 given in the table are equivalent whether velocity, net applied water, or depth of leaching are considered, and these values can be compared in different studies. It is significant that studies 4 and 5 [49,36, respectively] observed considerably different values of σ^2 and were conducted on the same field. The larger value of σ^2 found in the solution sampler study 4 could have been caused partly by transient conditions resulting from erratic rainfall [49] which can cause phase lags between net applied water and drainage rate at the solution sampler depth.

IMPLICATIONS OF SPATIAL VARIABILITY ON THE MEASUREMENT PROCESS

Table 11-5 gives sample sizes required in order to have at least a 95 percent probability of detecting a relative change of 20, 40, and 100 percent in the value of the mean when using a one-sample two-tailed t-test with probability of Type 1 error set at α=5 percent. These values were obtained using an approximation formula [50] which states that to have probability (1-β) of detecting a shift of F percent in the value of the mean using a one-sample two-tailed t-test with probability of Type 1 error set at α, one needs

$$N = \{(Z_{\alpha/2} + Z_\beta)^2(CV)^2/F^2\} + 0.5(Z_{\alpha/2})^2 \qquad (11\text{-}1)$$

where Z_β is the upper 100β percentile of the standard normal distribution and $Z_{\alpha/2}$ is similarly defined. The value of N given by

Table 11-4. Log Distribution of Solute Transport Velocity Parameters in Field Experiments.

Parameter	Soil	Area	Water Application	Measurements	N	σ^2	Reference
1. Velocity	loamy sand	4.6 x 6.1m^2 (4 plots)	ponding	solution samples to 240 cm	44	.48	[46]
2. Velocity	clay	8m x 8m	tricklers	solution samples to 150 cm	24	.32	[47]
3. Velocity	loam	150 (20 plots)	ponding	solution samples to 180 cm	120	1.56	[48]
4. Net applied water	sand	.64 ha (14 sites)	rainfall	solution samples to 180 cm	70	.32	[49]
5. Peak height	sand	.64 ha (36 sites)	sprinkler	soil cores to 300 cm	36	.12	[36]
6. Peak height	loamy sand	3m x 3m (8 plots)	sprinkler	soil cores to 80 cm	32	.45	[38]

Table 11-5. Sample Sizes Required to Have a 95% Probability of Detecting a Change of F% in the Mean Using a T-Test with α=5% (Sample CV'S are the mean of all field studies).

Parameter	Number of Studies	F: 20%	40%	100%	Average CV±SD
		Number of Samples			
Bulk density or porosity	13	6	*	*	10 ± 6
Percent sand or clay	10	28	9	*	28 ± 18
0.1 bar water content	4	9	*	*	14 ± 7
15 bar water content	5	23	7	*	25 ± 14
pH	4	4	*	*	8 ± 5
Saturated K	13	502	127	22	124 ± 71
Infiltration rate	8	135	36	8	64 ± 26
K_0 in $K(\theta)$	4	997	251	42	175 ± 139
Ponded solute velocity	1	1,225	308	51	194
Unsaturated solute velocity	5	127	33	7	62 ± 9
Solute concentration	4	119-551	32-140	7-24	60-130 (range)

*Sample size estimates are less than 5 and should not be used.

the formula is not usually an integer, and to obtain the sample size one should round up to the next higher integer. As an example, consider the case where the CV is 10 percent and one wants to have a 95 percent chance of detecting a 20 percent change in the mean using a t-test with α=5 percent, then

$$N = (1.96 + 1.645)^2(10)^2/(20)^2 + 0.5(1.96)^2 = 5.17$$

Hence one would want to use at least 6 measurements in this case (in using this approximation formula one always rounds up to the next highest integer).

Table 11-5 presents calculations of sample numbers using equation (11-1) required to estimate parameter mean values within F = 0.5, 0.2, or 0.1 of the true population mean at 5 percent precision using coefficient of variation values averaged over all the studies summarized above. These numbers could be thought of as rough guidelines for sampling information content based on the experimental information available.

The large sample numbers required for transport properties such as saturated hydraulic conductivity seem at first glance to pose an almost insurmountable barrier for field experimentation. However, it should be kept in mind that the arithmetic average hydraulic conductivity for an entire field is not necessarily a useful parameter. It may be more useful to obtain an estimate of the variation in hydraulic conductivity and use this information to construct a crude sample frequency diagram. Second, it may not always be necessary to construct highly accurate representations of the field average particularly if best- and worst-case scenarios are used. In this case, one might make a limited number of measurements and then run sensitivity analyses for ±100 percent variation of the parameters; this would seem reasonable given the experimental information summarized on the coefficient of variability of this parameter.

Finally, significant research is being conducted at this moment trying to connect the extensive variability information available through sources such as the U.S. Soil Conservation Survey and values of parameters such as those described above. Several promising efforts in this regard have already been accomplished. Duffy et al. [32], used Soil Survey descriptions to predict values for hydraulic conductivity within well-described mapping units. Comparison of these predictions with measurements inside these mapping units produced a correlation of $r^2 = 0.93$. Alexander [51] reviewed 720 measurements of wet bulk density at 1/3 bar applied suction which were measured on alluvial and upland soils covering 7 soil orders. The measured mean values were tested against multiple regression on a large number of soil properties available through Soil Survey information. For both the upland and alluvial soils a strong correlation was found between bulk density and functions of organic carbon and 15 bar water content. Cosby et al. [52], analyzed over 1400 measurements from Agricultural Research Service reports and found significant linear relationships between midpoint particle size fractions of the 12 textural groups of the USDA classification and each of the following quantities: mean of the field porosity, standard deviation of the porosity, mean of the log of saturated hydraulic conductivity, and standard deviation of the log of saturated hydraulic conductivity. In addition, they fitted measured moisture retention data to the model given in equation (11-2).

$$\Psi(\theta) = \Psi_2 [\theta/\theta_2]^b \qquad (11\text{-}2)$$

where Ψ_2 and b are constant parameters. They also found significant linear relationships between particle size and log Ψ_s, b and standard deviation of b. Results of this type could assist greatly in providing initial parameter estimates for areas where only soil survey information is available. Further study is necessary to determine the generality of these relationships found in the individual studies and also to establish guidelines for the precision of the correlation estimates.

STATISTICAL MODELS OF VARIABILITY

Statistical models for spatially variable quantities have been studied only in soil science disciplines for a few years. Thus, the information summarized below is far from conclusive. Since the potential benefits for using a statistical model to interpret spatially variable phenomena are high, future research into both scaling theory and regionalized variable analysis are strongly recommended.

Scaling Theory

Scaling theory, which is a form of similitude analysis, manipulates the mathematical equations describing the transport process and forms dimensionless groups with parameters chosen either on the basis of physical intuition or by making specific assumptions about the geometric relationships between different parts of the porous medium [10]. The latter philosophy, called similar media analysis, treats the porous medium as a scale heterogeneous body consisting locally of magnified versions of a reference porous medium in which all parts of the geometry are magnified by the same scale relation relative to this standard [10]. This geometric relationship to the standard porous medium can thus be represented by a single scaling factor λ giving the ratio of the length of a unit cell (e.g., a pore) of the porous medium relative to the length of the same unit cell in the reference porous medium. Thus, the scale factor may be thought of as a magnification factor. If the scaling theory hypothesis is reasonably well satisfied for a field, then a number of simplifications for describing spatial variability result immediately. First, as shown by Warrick and Amoozegar-Fard [11], the standard water transport equation may be scaled and thus needs to be be solved only once to represent a simulation of any one-dimensional transport process within the porous medium once the promedium is known. Since this microscopic scaling factor cannot be measured directly, it must be inferred from measurements of soil physical parameters such as those described above. These parameters are then compared to the reference site measurement of the parameter in order to yield the scaling factor. The relationships between various measurable parameters and the corresponding scaling factor relationships are reviewed in Miller [53].

Table 11-6 presents published parameters of the scaling factor distributions which were inferred from various experiments which measured soil physical properties randomly or systematically across a field unit. In each of these studies, the scaling factor values

Table 11-6. Lognormal Scaling Factor Parameters Measured in Field Experiments.

Soil Type	Field Size	M(Lnλ)	S^2(Lnλ)	CV(%)	Measure-ments	Scaling Method*	Reference
Panoche	150 ha	-.13	0.51	55	120	A	[8]
silty		-.63	1.17	170	120	B	[8]
clay		-.14	0.59	64	20	C	[15]
loam		-.13	0.50	53	120	D	[15]
Pima loam	87 ha	-.10	0.45	48	180	A	[8]
Teller	<7.2 km	-.45	1.04	139	64	A	[8]
Silt loam	9.6 ha	-.46	1.06	143	26	E	[29]
		-.05	0.33	34	26	F	[29]
Hamra sand	0.8 ha	-.05	0.30	31	120	A	[9]
		-.13	0.51	55	120	B	[9]
		-.16	0.57	62	120	G	[9]
		-.08	0.41	43	120	H	[9]
Yolo loam	0.7 ha	-.38	0.77	90	72	A	[35]

*A Scale matric potential - saturation curve
B Scale hydraulic conductivity - saturation curve
C Scale steady state infiltration
D Calculate from lognormal saturated conductivity distribution
E Scale S in Philip infiltration equation
F Scale A in Philip infiltration equation
G Scale wetting front distance (Method 1)
H Scale wetting front distance (Method 2)

were fitted to log-normal distributions. Furthermore, all scale factors have been normalized so that they have a sample mean value equal to 1.0 while assigning a reference value of unity to the reference location of the porous medium.

Immediately obvious from this table is the fact that the distribution of parameters is not related to soil type or field size. However, there is apparently a relationship between the variance of the scaling factor distribution and the type of property measurement used to estimate the scaling factors. In fact, the first four values of the scaling factor distribution given on Table 11-7 were all obtained from an analysis of the same redistribution experiment published by Nielsen et al. [15]. The first two scaling factor distribution parameters were obtained in the study by Warrick et al. [8], who scaled matric potential and hydraulic conductivity, respectively, to form two distributions of the scaling factor.

Table 11-7. Correlation Length Parameters Measured in Field Experiments.

Parameter	Soil	Field Size (ha)	Sample Spacing (m)	Correlation Length (m)	Reference
Saturated K	sand	0.8	$\sim$1.0	25	[16]
	-	-	1.0	1.6	[55]
Infiltration rate	clay loam	0.004	.05	.13	[31]
	weathered	0.08	2.0	<2.0	[56]
	shale sandy loam	0.9	1.0	35	[30]
Percent sand	sandy clay	0.3	10	36	[25]
	loam sand	7.2	10	30	[57]
pH	-	-	16	196	[58]
	variable	Hawaii	1,500	2,3000	[59]
	clay loam	85	0.2	1.5	[24]
	clay loam	85	2.0	21.5	[24]
	clay loam	85	20.0	130	[24]
EC	clay laom	455	80	800	[40]
	clay loam	85	0.2	1.2	[24]
	clay loam	85	2.0	20.0	[24]
	clay loam	85	20.0	20.0	[24]

According to the similar medium hypothesis, these two distributions should have been identical. Instead, the variance of the hydraulic conductivity scaling factors was substantially higher than that of the matric potential scaling factor distribution. Although a rather high correlation ($r^2 = 0.85$) was observed between the two distributions, the extra spreading found in the hydraulic conductivity distribution was undoubtedly due to the higher error involved in this measurement. The third set of values from Table 11-6 was obtained by directly scaling the 20 steady state infiltration values on the plots from the Nielsen et al. [15], study prior to the redistribution experiment. The fourth value was obtained by scaling the unsaturated hydraulic conductivity-water content relationship for these field plots as presented in the study of Libardi et al. [33]. If this field behaves as a similar media, and if each of the measurements have the same relative precision, the scaling factor parameter should have been described by the same distribution. The substantially different values found in Table 11-6 are evidence of the failure of the similar media hypothesis for this field.

A similar lack of correspondence between different methods of measuring the scaling factor parameters was observed by Sharma et al. [29], who scaled the S and A parameters in the Philip infiltration equation according to the similar medium hypothesis. Although the model itself described infiltration remarkably well at each of their 24 sites, the inferred scaling factor distribution using those equations produced two distributions with widely differing properties. As in the case of the Warrick et al. [8], study, these two parameter distributions were highly correlated but represented distributions with quite different variance. A third evidence of the failure of the similar media hypothesis was found in the study of Russo and Bresler [16] in which four methods were used to determine the scaling factor distributions. As in the case of the Warrick study, hydraulic conductivity scaling produced a higher variance than the matric potential scaling method even though the parameters estimated by the two methods were correlated.

Although the studies conducted above indicate that the similar medium hypothesis is not a good conceptual model for describing soil variability, this represents but one type of scaling possibility to use for field heterogeneity. Other options include separately scaling the hydraulic conductivity and matric potential distribution as was done by Simmons et al. [35], or even by using kinematic or dynamic scaling which may involve scaling of boundary conditions as well [54]. In the latter philosophy, scaling factors may be developed for a particular type of process (e.g. infiltration) and the scaling factors represented from this would be applicable to a set of similar processes. Since a major advantage of scaling theory would be a reduction in the numbers of measurements, this general approach is deserving of further study.

Regionalized Variable Analysis

An alternate approach for describing a spatially variable phenomena has been to apply the theory of regionalized variables to the interpretation of discrete measurements of spatially variable soil physical properties. In this approach, a value of the parameter measured at a particular location is regarded as a member of a statistical set of such values described by probability laws. In order to apply the theory, it is necessary to assume that the property is stationary in space so that each location is described by the same probability distribution and spatial covariances depend only on the separation between measurements and not on the absolute location of the measurement itself [6]. Thus, when a set of N measurements are taken on a field, an analysis is made not only of the sample mean and sample variance, but also on the spatial correlations between measurements as a function of the distance between the measurements. This is accomplished in the following way. An ordered spatial sampling pattern, usually a transect in one dimension or a grid in two dimensions, is set up across the field. A special quantity called the semi-variance is calculated as the mean square deviation of all replicate measurements a given distance apart.

$$\gamma^2[h] = \frac{1}{2N(h)} \sum_{J=1}^{N} [Z(x+h)-Z(x)]^2 \qquad (11\text{-}3)$$

where N(h) is the number of replicates separated by a lag distance h, and Z is the value of the parameter. If this semi-variance increases to a plateau when plotted as a function of lag distance or a separation, the physical interpretation is made that the parameter is spatially correlated with like measurements of itself within a zone defined by the distance required to reach this plateau. The plateau distance which may be measured by this method or by similar statistical procedures is variously called the zone of correlation, the integral scale, or the range. A review of methods used to calculate the length is given in Peck [3].

Analysis of spatial correlations offers a potentially powerful tool both for sample design and for finding the optimum method of analyzing the information yielded by a finite set of samples on a field. For example, if it were known that the correlation length of hydraulic conductivity were 20 meters, then samples spaced at precisely this distance would provide information with no redundancy caused by overlap.

Table 11-7 summarizes the correlation parameters which have been inferred from experimental studies of the spatial correlation of soil physical properties. The small number of studies presented reflects both the relatively short period of time that such procedures have been in operation and also the difficulty and expense involved in constructing experiments on the scale required to yield such information. The four parameters indexed in this table represent different ways of interpreting the spatial dependence parameters but are highly correlated with each other for a given study and for all practical purposes may be considered equivalent.

The information summarized in this table is rather ambiguous. Although the static soil properties are generally correlated over greater distance than the dynamic properties, variations of the correlation length of all properties between studies seem to be much more substantial than the variations of a given property between studies. Significantly, there is a strong correlation between the inferred zone of influence and distance between discrete samples used to construct the spatial correlation function. For example, the pH values taken from the study of Gajem et al. [24], in Table 11-7 show an apparent correlation length of 1.5 meters when a 20 cm transect spacing was used, a correlation of 21.6 meters when a 2 meter spacing was used, and a correlation of 130 meters when a spacing of 20 meters was used all on the same field. Furthermore, the total variance of pH was similar in all three cases.

This ambiguity becomes quite obvious when studies conducted at dramatically different sampling densities are compared. For example, the pH measurements taken by Yost et al. [59], using a sample spacing of 800 meters were found to be correlated up to 2 kilometers whereas the same chemical parameter was apparently not correlated at distances greater than 1.5 meters in the Gajem study.

There are various theoretical explanations offered for this so-called scale effect. It is to be expected that the variance of a property will change with the scale of observation so that larger spacing may involve detection of a larger scale of correlation. However, in such a case, the total variance of the sample set should be larger than the variance of a sample set analyzed within a smaller scale. An obvious way in which this scale effect can come about is by a violation of the stationarity hypothesis. If the underlying mean value of the parameter has a functional drift with distance, this can bias the interpretation of the underlying stochastic component in a manner similar to that discussed above [60]. Drift removal, however, presents a particularly difficult problem since the functional form of the drift function is not initially known. At the present time, only somewhat subjective procedures for drift removal have been recommended; for example, an interactive procedure in which candidate drift functions are subtracted from the original sample set, and the statistical analysis is repeated until some optimum criteria is satisfied [6].

Regionalized variable analysis is at the present time a science in its infancy. Future investigations will undoubtedly develop some objective statistical criteria for making the stochastic measurements and interpreting the results which will remove much of the ambiguity found in the present study. However, it should be cautioned that the zone of correlation or its equivalent is a required property of several stochastic transport models [12,13]. If these properties cannot be objectively measured, then the predictive capability of the stochastic models becomes doubtful.

ACKNOWLEDGEMENT

Appreciation is expressed to Dr. Thomas H. Starks for assistance in calculating Table 11-5.

REFERENCES

1. Jury, W. A. "Simulation of Solute Transport with a Transfer Function Model," Water Resour. Res. 18:363-368 (1982).

2. Warrick, A. W., and D. R. Nielsen. Applications of Soil Physics (New York: Academic Press, 1980).

3. Peck, A. J. "Field Variability of Soil Physical Properties," Adv. in Irrig. 2:189-221 (1984).

4. Vieira, S. R., J. L. Hatfield, D. R. Nielsen, and J. W. Biggar. "Geostatistical Theory and Application to Variability of Some Agronomical Properties," Hilgardia 51:1-75 (1983).

5. McBratney, A. B., and R. Webster. "How Many Observations are Needed for Regional Estimation of Soil Properties," Soil Sci. 135:177-183 (1983).

6. Journel, A. G., and C. J. Huijbregts. Mining Geostatistics (London: Academic Press, 1978).

7. Peck, A. J., R. J. Luxmoore, and L. J. Stolzy. "Effects of Spatial Variability of Soil Hydraulic Properties in Water Budget Modeling," Water Resour. Res. 13:348-354 (1977).

8. Warrick, A. W., G. J. Mullen, and D. R. Nielsen. "Scaling Field Measured Soil Hydraulic Properties Using a Similar Media Concept," Water Resour. Res. 13:355-362 (1977).

9. Russo, D., and E. Bresler. "Scaling Soil Hydraulic Properties of a Heterogeneous Field," Soil Sci. Soc. Am. J. 44:681-683 (1980).

10. Miller, E. E., and R. D. Miller. "Physical Theory for Capillary Flow Phenomena," J. Appl. Phys. 27:324-332 (1956).

11. Warrick, A. W., and A. Amoozegar-Fard. "Infiltration and Drainage Calculations Using Spatially Scaled Hydraulic Properties," Water Resour. Res. 15:1116-1120 (1979).

12. Gelhar, L. W., A. L. Gutjahr, and R. L. Naff. "Stochastic Analysis of Macrodispersion in a Stratified Aquifer," Water Resour. Res. 15:1287-1397 (1979).

13. Gelhar, L. W., and C. Axness. "Three Dimensional Stochastic Analysis of Macrodispersion in Aquifers," Water Resour. Res. 19:161-180 (1983).

14. Jury, W. A. "A Review of Published Studies of Field Measured Soil Water and Chemical Properties," EPRI Topical Report (In press).

15. Nielsen, D. R., J. W. Biggar, and K. T. Erh. "Spatial Variability of Field Measured Soil Water Properties," Hilgardia 42:215-259 (1973).

16. Russo, D., and E. Bresler. "Soil Hydraulic Properties as Stochastic Processes: 1. Analysis of Field Spatial Variability," Soil Sci. Soc. Amer. J. 45:682-687 (1981).

17. Cameron, D. R. "Variability of Soil Water Retention Curves and Predicted Hydraulic Conductivities," Soil Sci. 126:364-371 (1978).

18. Cassel, D. K. "Spatial and Temporal Variability of Soil Physical Properties Following Tillage of Norfolk Loamy Sand," Soil Sci. Soc. Amer. J. 47:196-201 (1983).

19. Gumaa, G. A. "Spatial Variability of In Situ Available Water," Ph.D. Thesis, University of Arizona, Tucson (Available as 78-24365 from Xerox University Microfilms, Ann Arbor, MI) (1978).

20. Courtin, P., M. C. Feller, and K. Klinka. "Lateral Variability in Some Properties of Disturbed Forest Soils in S.W. British Columbia," Canad. J. Soil Sci. 63:529-539 (1983).

21. Cassel, D. K., and A. Bauer. "Spatial Variability in Soils Below Depth of Tillage: Bulk Density and Fifteen Atmosphere Percentage," Soil Sci. Soc. Amer. Proc. 39:247-250 (1975).

22. Babalola, O. "Spatial Variability of Soil Water Properties in Tropical Soils of Nigeria," Soil Sci. 126:269-279 (1978).

23. Stockton, J. G., and A. W. Warrick. "Spatial Variability of Unsaturated Hydraulic Conductivity," Soil Sci. Soc. Amer. J. 36:847-849 (1971).

24. Gajem, Y. M., A. W. Warrick, and D. E. Myers. "Spatial Dependence of Physical Properties of a Typic Torrifluvent Soil," Soil Sci. Soc. Amer. J. 45:709-715 (1981).

25. Vauclin, M., S. R. Vieira, G. Vachaud and D. R. Nielsen. "Use of Co-Kriging with Limited Field Soil Observations," Soil Sci. Soc. Amer. J. 47:175-184 (1983).

26. Cameron, D. R., M. Nyborg, J. A. Toogood, and D. H. Laverty. "Accuracy of Field Sampling for Soil Tests," Canad. J. Soil Sci. 51:166-175 (1971).

27. El Abd, H. "Spatial Variability of the Pesticide Distribution Coefficient," Ph.D. Thesis, University of California, Riverside (1984).

28. Willardson, L. S., and R. L. Hurst. "Sample Size Estimates in Permeability Studies," Amer. Soc. Civil Eng. 91:1-9 (1965).

29. Sharma, M. L., G. A. Gander, and C. G. Hunt. "Spatial Variability of Infiltration in a Watershed," J. Hydrol. 45:101-122 (1980).

30. Vieira, S. R., D. R. Nielsen, and J. W. Biggar. "Spatial Variability of Field-Measured Infiltration Rate," Soil Sci. Soc. Amer. J. 45:1040-1048 (1981).

31. Sisson, J. B., and P. J. Wierenga. "Spatial Variability of Steady State Infiltration Rates as a Stochastic Process," Soil Sci. Soc. Amer. J. 45:699-704 (1981).

32. Duffy, C., P. J. Wierenga, and R. A. Kselik. "Variations in Infiltration Rate Based on Soil Survey Information and Field Measurements," Agriculture Experimental Station Bulletin 60, New Mexico State University (1981).

33. Libardi, P. L., K. Reichardt, D. R. Nielsen, and J. W. Biggar. "Simple Field Methods for Estimating Soil Hydraulic Conductivity," Soil Sci. Soc. Amer. J. 44:3-6 (1980).

34. Wagenet, R. J., and B. K. Rao. "Description of Nitrogen Movement in the Presence of Spatially Variable Soil Hydraulic Properties," Agr. Water Mgmt. 6:227-243 (1983).

35. Simmons, C. S., D. R. Nielsen, and J. W. Biggar. "Scaling of Field-Measured Soil-Water Properties," Hilgardia 47:77-174 (1979).

36. Jury, W. A., H. El Abd and T. M. Collins. "Field Scale Transport of Nonadsorbing and Adsorbing Chemicals Applied to the Soil Surface," Proceedings of the NWWA/U.S. EPA Conference on Characterization and Monitoring of the Vadose (Unsaturated) Zone, Las Vegas, NV (1983).

37. Richter, G. "Microlysimeter and Field Study of Water and Chemical Movement Through Soil," M.S. Thesis, University of California, Riverside (1984).

38. Wild, A., and I. A. Babiker. "The Asymmetric Leaching Pattern of NO_3 and Cl in a Loamy Sand Under Field Conditions," J. Soil Sci. 27:460-466 (1976).

39. Wagenet, R. J., and J. J. Jurinak. "Spatial Variability of Solute Salt Content in a Mankos Shale Watershed," Soil Sci. 126:342-349 (1978).

40. Hajrasuliha, S., N. Baniabbassi, J. Metthey, and D. R. Nielsen. "Spatial Variability of Soil Sampling for Salinity Studies in Southwest Iran," Irrig. Sci. 1:197-108 (1980).

41. Lund, L. J. "Variations in NO_3 and Cl Concentrations Below Selected Agricultrual Fields. Soil Sci. Soc. Amer. J. 46: 1062-1067 (1982).

42. Johnston, C. D., W. M. McArthur and A. J. Peck. "Distribution of Soluble Salts in Soils of the Manjimup Woodchip License Area, Western Australia," Land Res. Mgmt. Tech. Paper 5, CSIRO, Australia (1980).

43. Pratt, P. F., W. W. Jones, and V. E. Hunsaker. "Nitrate in Deep Soil Profiles in Relation to Fertilizer Rates and Leaching Volume," J. Environ. Qual. 1:97-102 (1972).

44. Oster, J. D., and J. D. Wood. "Hydrosalinity Models: Sensitivity to Input Variables," in Proceedings of the National Conference on Irrigation Return Flow Quality Management, Colorado State University (1977).

45. Jury, W. A. "Chemical Transport Modeling: Current Research and Unresolved Problems," in Chemical Mobility and Reactivity in Soil Systems, Special Publication II (Soil Science Society of America, 1983).

46. Starr, J. L., H. C. DeRoo, C. R. Frink and J. Y. Parlange. "Leaching Characteristics of Layered Field Soil," Soil Sci. Soc. Amer. J. 42:386-391 (1978).

47. Van de Pol, R. M. "Solute Movement in a Field Soil," M.S. Thesis, New Mexico State University, Las Cruces (1974).

48. Biggar, J. W., and D. R. Nielsen. "Spatial Variability of the Leaching Characteristics of a Field Soil," Water Resour. Res. 12:78-84 (1976).

49. Jury, W. A., L. H. Stolzy, and P. Shouse. "A Field Test of the Transfer Function Model for Predicting Solute Transport," Water Resour. Res. 18:369-375 (1982).

50. Guenther, W. C. "Sample Size Formulas for Normal Theory T-Tests," American Statistician 35:243-244 (1981).

51. Alexander, E. B. "Bulk Densities of California Soils in Relation to Other Soil Properties," Soil Sci. Soc. Amer. J. 44:689-692 (1980).

52. Cosby, B. J., G. M. Hornberger, R. B. Clapp, and T. R. Ginn. "A Statistical Exploration of the Relationships of Soil Moisture Characteristics to the Physical Properties of Soils," Water Resour. Res. 6:682-690 (1984).

53. Miller, E. E. "Similitude and Scaling of Soil Water Phenomena," in Applications of Soil Physics (New York: Academic Press, 1980).

54. Tillotson, P., and D. R. Nielsen. "Scale Factors in Soil Science," Soil Sci. Soc. Amer. J. 48:953-959 (1984).

55. Gelhar, L. W., A. A. Baker, A. L. Gutjahr, and J. R. MacMillan. "Comments on a Stochastic-Conceptual Analysis of One Dimensional Groundwater Flow in Nonuniform Homogeneous Media by R. A. Freeze," Water Resour. Res. 13:477-479 (1977).

56. Luxmoore, R. J., B. P. Spalding, and I. M. Munro. "Areal Variation and Themical Modification of Weathered Shale Infiltration Characteristics," Soil Sci. Soc. Amer. J. 45:687-691 (1981).

57. Campbell, J. B. "Spatial Variation of Sand Content and pH Within Single Continuous Delineations of Two Soil Mapping Units," Soil Sci. Soc. Amer. J. 42:460-464 (1978).

58. McBratney, A. B., and R. Webster. "Spatial Dependence and Classification of the Soil Along a Transect in Northeast Scotland," Geoderma 26:63-82 (1981).

59. Yost, R. S., G. Uehara, and R. L. Fox. "Geostatistical Analysis of Soil Chemical Properties of Large Land Areas," Soil Sci. Soc. Amer. J. 46:1028-1032 (1982).

60. Starks, T. H., and J. H. Fang. "The Effect of Drift on the Experimental Semivariogram," Mathematical Geol. 14:309-311 (1982).

APPENDIX A

MATHEMATICAL DERIVATION OF CHEMICAL TRANSPORT EQUATIONS

W. A. Jury
University of California-Riverside, Riverside, California

CONTINUITY EQUATIONS

Conservation Law

The differential equations describing water, heat, or chemical transport through porous media are macroscopic and make no attempt to describe three-dimensional flow at the scale of a soil pore or soil particle. Equations are written and averaged over a volume element large enough to produce a meaningful average value which is independent of the properties of the averaging volume for such quantities as volumetric water content, bulk density, etc. This scale, often called the Darcy scale, is itself a function of the degree of heterogeneity of the porous medium [1]. As a simplification, the equations derived below will be one-dimensional with the direction assumed to represent the vertical coordinate z. Although in principle, balance equations can be written for three dimensions, in practice very little use has been made of multi-dimensional transport equations except under very restrictive assumptions. The one-dimensional equations derived here will be appropriate for simulation of many practical transport processes such as volatilization and vertical chemical transport by leaching.

The mass balance of water, assumed to be flowing in the z direction, is achieved by looking at a small cubic soil volume element of thickness Δz and unit cross-sectional area $\Delta x \Delta y = 1$ for which the conceptual balance equation given in equation (A-1) is written.

[rate of water flow into V] - [rate of water flow out of V]

= [rate of increase of water stored in V]

+ [rate of water uptake in V by plant roots] (A-1)

where V is the soil volume.

If we represent the volumetric water flux by J_w ($cm^3/cm^2/day$) and the water storage per unit volume by the volumetric water content θ (cm^3/cm^3) then we may rewrite equation (A-1) as

$$J_w(z)\Delta x\Delta y - J_w(z+\Delta z)\Delta x\Delta y = \frac{\Delta\theta}{\Delta t}\Delta x\Delta y\Delta z + S_w\Delta x\Delta y\Delta z \quad \text{(A-2)}$$

where S_w (day^{-1}) is the rate of water uptake per unit soil volume.

If we divide equation (A-2) through by the volume element $\Delta x\Delta y\Delta z$ and take the limit of small time and small volume, we obtain the water conservation equation (A-3)

$$\frac{\partial J_w}{\partial z} + \frac{\partial\theta}{\partial t} S_w = 0 \quad \text{(A-3)}$$

Chemical Conservation Law

In a similar manner, we may write a conceptual equation for any chemical species of interest as

[rate of flow of chemical into V] -

[rate of flow of chemical out of V]

= [rate of increase of chemical stored in V]

+ [rate of disappearance of chemical from V by reactions] (A-4)

If we represent the chemical transport flux as J_s ($g/cm^2/day$) and the mass of chemical stored per soil volume (i.e., the concentration) as C_T (g/cm^3), we may rewrite equation (A-4) as

$$J_s(z)\Delta x\Delta y - J_s(z+\Delta z)\Delta x\Delta y = \frac{\Delta C_T}{\Delta t}\Delta x\Delta y\Delta z + S_s\Delta x\Delta y\Delta z \quad \text{(A-5)}$$

where S_s ($g/cm^3/day$) is the term representing the rate of disappearance of chemical per soil volume by chemical or biological reactions.

If we again divide equation (A-5) through by the volume element $\Delta x \Delta y \Delta z$ and take the limit of small volume and small time, we obtain the chemical species balance equation (A-6)

$$\frac{\partial C_T}{\partial t} + \frac{\partial J_S}{\partial z} + S_S = 0 \qquad \text{(A-6)}$$

FLUX EQUATIONS

Water Flux J_W

The one-dimensional flux of water through unsaturated soil is given by the Buckingham-Darcy flux law [2]

$$J_W = -K(h) \left[\frac{\partial h}{\partial z} + 1\right] \qquad \text{(A-7)}$$

where h(cm) is matric potential head and K(h) (cm/day) is the unsaturated hydraulic conductivity; this is a very non-linear function of the matric potential head and may decrease up to seven orders of magnitude between very moist ($h \rightarrow 0$) and very dry ($h \rightarrow -\infty$) soil [2].

Chemical Flux J_S

Diffusion Fluxes

Vapor diffusion. The vapor diffusion flux in soil J_G (g/cm^2/day) is given by the extended form of Fick's law of diffusion [3]

$$J_G = \eta_G (a) D^{AIR}_G \frac{\partial C_G}{\partial z} = - D^{SOIL}_G \frac{\partial C_G}{\partial z} \qquad \text{(A-8)}$$

when D^{AIR}_G is the gaseous diffusion coefficient of the chemical species in free air, C_G (g/cm^3 soil air) is the mass of chemical vapor per volume of soil air, a is volumetric air content (cm^3/cm^3), and η_G is the tortuosity factor. This tortuosity factor accounts for the decreased cross-sectional area and increased path length induced by a porous medium, which is a function of the volumetric air space available for transport [4].

There have been numerous structural models introduced to represent the tortuosity factor η in soil. The formulation which has proved to be the most useful over a large range of air content is the Millington and Quirk formulation

$$\eta_G(a) = a^{10/3}/\emptyset^2 \tag{A-9}$$

[5] where $\emptyset$ is soil porosity. This model has proven to be very successful at representing the tortuosity effect on diffusion coefficients of pesticides in soil [6,7,8,9] even at water contents very close to saturation [10].

Liquid Diffusion. In analogy to the above diffusion model, the liquid diffusion transport flux J_L ($g/cm^2/day$) is given by equation (A-10)

$$J_L = -\eta_L(\theta)\, D^{WATER}{}_L \frac{\partial C_L}{\partial z} = -D^{SOIL}{}_L \frac{\partial C_L}{\partial z} \tag{A-10}$$

where C_L (g/cm^3 solution) is mass of liquid per volume of soil solution, $D^{WATER}{}_L$ (cm^2/day) is the molecular diffusion coefficient of the chemical species in pure water, and η_L is the tortuosity factor accounting for the decreased cross-sectional area and increased path length for molecular diffusion in solution due to the porous medium. Research studies have also validated the use of the Millington and Quirk [5] equation for this tortuosity factor [7,9]. The Millington and Quirk model for liquid water is achieved by replacing the air content a in equation (A-9) by the water content θ

$$\eta_L(\theta) = \theta^{10/3}/\emptyset^2 \tag{A-11}$$

Dispersion Flux

Chemical transport equations must include a term to account for hydrodynamic dispersion since pore-scale mass fluxes are replaced by a volume-averaged mass flux. This dispersion flux J_{HD} for one-dimensional flow is given by equation (A-12)

$$J_{HD} = -D_{HD} \frac{\partial C_L}{\partial z} \tag{A-12}$$

where D_{HD} is the hydrodynamic dispersion coefficient. The dependence of D_{HD} on soil and environmental properties is not well understood at present, but numerous studies have shown that D_{HD} increases as water flux J_w increases. In most cases, this dependence is assumed to be linear, as in equation (A-13)

$$D_{HD} = d_{HD}\, J_w \tag{A-13}$$

where d_{HD} is called the dispersivity [1]. In three-dimensional flow, D_{HD} is a tensor which is usually modeled as having longitudinal and transverse components [11].

Because of the similarity in form, the liquid diffusive and hydrodynamic dispersive fluxes equations (A-10) and (A-12) are usually lumped together as a combined flux J_{LHD}

$$J_{LHD} = J_{HD} + J_L = - D_{HD} + D^{SOIL}{}_L \frac{\partial C_L}{\partial z} = - D^{SOIL}{}_E \frac{\partial C_L}{\partial z} \qquad \text{(A-14)}$$

where $D^{SOIL}{}_E$ is the effective diffusion-dispersion coefficient. This formulation will be used in subsequent analysis.

Mass Flux

When water flow is occurring, substantial amounts of dissolved chemical can be transported with the moving soil solution. This transport flux J_M (cm^2/day) is given by equation (A-15)

$$J_M = C_L J_W \qquad \text{(A-15)}$$

Solute Flux

The combined effects of liquid and vapor diffusion and mass flow are summarized in the solute flux equation (A-16)

$$J_s = J_G + J_{LHD} + J_M = - D^{SOIL}{}_G \frac{\partial C_G}{\partial z} - D^{SOIL}{}_E \frac{\partial C_L}{\partial z} + C_L J_W \qquad \text{(A-16)}$$

STORAGE EQUATIONS

Since the amount of water adsorbed on soil minerals is constant until the soil reaches an extreme state of dryness and because the amount of water vapor stored is negligible, the water content per soil volume is adequately represented by the liquid water content per soil volume. In the case of chemicals, however, particularly pesticides and other trace organics, substantial amounts of material may be stored in all three phases. Thus, the total concentration per soil volume C_T in equation (A-6) is expanded to give the contributions in each phase

$$C_T = \rho_b C_s + \theta C_L + aC_G \qquad \text{(A-17)}$$

where C_s (g/g soil) is mass of chemical adsorbed per mass of soil and ρ_b is soil bulk density. These units are conventional and correspond to the usual method of measurement of the concentrations in each phase.

TRANSPORT EQUATIONS

Water Transport

When the water transport flux equation (A-7) is plugged into the continuity equation (A-3), the result is equation (A-18).

$$\frac{\partial \theta}{\partial t} + S_W = \frac{\partial}{\partial z}\left[K(h)\left(\frac{\partial h}{\partial z} + 1\right)\right] \qquad \text{(A-18)}$$

In its present form, this cannot be solved because there are two dependent variables: the matric potential head h and the volumetric water content θ. The equilibrium relationship $h(\theta)$ used to relate these two variables is called the water characteristic function or the matric potential-water content relation. This function is obtained experimentally. A problem, currently unresolved in research studies, is that a significant amount of hysteresis appears between the wetting curve and the drying curve of this relationship $h(\theta)$. For this reason, experiments and simulations using equation (A-18) have generally been monotonic wetting or drying studies. The water uptake function S_W is also determined experimentally. The experimental determination is usual only for fully developed crop root systems operating under a characteristic water regime, and, as a result this function does not change appreciably with time. For studies when no plant roots are present, the water flow equation may be written as

$$C(h)\frac{\partial h}{\partial t} = \frac{\partial}{\partial z}\left[K(h)\left(\frac{\partial h}{\partial z} + 1\right)\right] \qquad \text{(A-19)}$$

where

$$C(h) = \partial\theta/\partial h \qquad \text{(A-20)}$$

is the water capacity function which is the slope of the water content matric potential function for a monotonic process.

Equation (A-19) may also be recast with the volumetric water content as the dependent variable if this is more convenient [12].

Solute Transport

When the solute transport flux equation (A-16) and storage equation (A-17) are plugged into the continuity equation (A-6), the result is equation (A-21)

$$\frac{\partial}{\partial t}(\rho_b C_S + \theta C_L + aC_G) + S_S = \frac{\partial}{\partial z}\left[D^{SOIL}{}_E \frac{\partial C_L}{\partial z} - J_W C_L\right] \qquad \text{(A-21)}$$

A number of simplifications are needed in equation (A-21) before solution is possible, since there are three dependent variables C_G, C_L, C_S.

The relationship which is usually assumed to apply between the liquid concentration and the gaseous concentration is given by Henry's Law

$$C_G = K_H C_L \tag{A-22}$$

where K_H is Henry's constant. The applicability of Henry's Law to soil systems has been confirmed for a variety of chemicals and circumstances [13,14,15,16]. Furthermore, Spencer and Cliath [16] showed that the relationship equation (A-22) persisted from very trace concentration levels all the way to saturation levels of the organic chemicals they studied. Thus, one may estimate Henry's constant K_H as the ratio of saturated vapor density to aqueous solubility [16].

The relationship which exists at equilibrium between adsorbed concentration and liquid concentration is called the adsorption isotherm.

$$C_S = f(C_L) \tag{A-23}$$

This relationship is obtained under equilibrium laboratory conditions. The functional form which has most commonly been used to represent this relationship for pesticides and other trace organics is the so-called Freundlich relationship [17].

$$C_S = K_F(C_L)^N \tag{A-24}$$

where K_F and N are constants. Karickhoff et al. [18] and Karickhoff [19] have shown that for low aqueous concentrations of weakly polar compounds, a linear isotherm

$$C_S = K_D C_L \tag{A-25}$$

may be substituted for the non-linear relationship where K_D is called the distribution coefficient. This representation is only an approximation which has been measured in the C_S - C_L relation for a number of compounds [20,21].

Under natural conditions, equilibrium may not be completely reached between the phases. For this reason, kinetic or rate-limited expressions, particularly in chemical leaching studies, have been used to represent the relation between C_S and C_L. Lindstrom et al. [22], Oddson et al. [23], and van Genuchten et al. [20], all used the rate-limited expression given in equation (A-26) to represent the change in adsorbed concentration

$$\frac{\partial C_S}{\partial t} = r\,[f(C_L) - C_S] \tag{A-26}$$

where r is a rate coefficient. Equation (A-26) assumes that the change in adsorbed concentration is proportional to the deviation from equilibrium. In practice, no model has ever been able to predict the form of the rate coefficient r for a given soil process, and it has to be measured from experimental breakthrough data. Recently, however, Rao et al. [24], have successfully modeled the dependence of this factor r on certain geometric and flow properties of the porous medium.

PARTITIONING COEFFICIENTS

The large number of parameters in equations (A-21) through (A-26) have led researchers to make simplifying assumptions in order to reduce the need for calibration and repeated measurements. Jury et al. [3], introduced partitioning coefficients for each phase by assuming that Henry's Law, equation (A-22), and the linear isotherm, equation (A-25), were valid and that equilibrium was achieved in a short enough period of time so that kinetic effects on adsorption could be neglected. With these two assumptions, the storage equation (A-17) may be written as

$$C_T = R_S C_S = R_L C_L = R_G C_G \tag{A-27}$$

where

$$R_S = C_T/C_S = \rho_b + \theta/K_D + aK_H/K_D \tag{A-28}$$

$$R_L = C_T/C_L = \rho_b K_D + \theta + aK_H \tag{A-29}$$

$$R_G = C_T/C_G = \rho_b K_D/K_H + \theta/K_H + a \tag{A-30}$$

are the adsorbed, liquid, and vapor partition coefficients, respectively. The inverse of the R coefficients gives the fraction of total concentration in each phase. The expressions in equations (A-27) through (A-30) may be used to make a number of simplifying definitions in the above equations. For example, the solute flux equation (A-16) may now be written in terms of total concentration

$$J_S = - D_E \frac{\partial C_T}{\partial z} + V_E C_T \tag{A-31}$$

where

$$D_E = (D^{SOIL}{}_G K_H + D^{SOIL}{}_E)/R_L \tag{A-32}$$

is the effective diffusion-dispersion coefficient for both liquid and vapor diffusion and

$$V_E = J_W/R_L \tag{A-33}$$

is the effective solute velocity which can be used to make estimates of residence time for a given chemical.

REACTION RATES

The generalized reaction term S_S in equation (A-21) is merely symbolic and for a given chemical species may depend on a variety of factors. For multi-electrolyte solutions, for example, the reaction term depends on chemical activity and thus involves all species in solution. Simple models of reaction such as precipitation and dissolution have been achieved by combining chemical equilibrium models with transport models [25]. For pesticides and other trace organics, reactions such as hydrolysis, photolysis, and microbial degradation have been modeled (see chapter on these processes), but within soil, all reactions are usually represented by an effective rate coefficient μ_E. Although this rate coefficient depends on such factors as water content, organic matter content, temperature, etc., it is generally represented as a first-order constant rate coefficient for simplicity [17,24,26]. The reason for this simplification is that most measurements of degradation are made by observing residual concentrations after a certain period of time and by fitting them to a model assuming exponential depletion. With the partitioning coefficient formulation, the first-order rate expression S_s is written as

$$S_s = \mu_E C_T \tag{A-34}$$

where μ_E is the effective rate coefficient for all phases combined. With these simplifications, equation (A-21) becomes

$$\frac{\partial C_T}{\partial t} = \frac{\partial}{\partial z}\left[D_E \frac{\partial C_T}{\partial z} - V_E C_T\right] - \mu_E C_T \tag{A-35}$$

This equation is the so-called convection-dispersion equation. van Genuchten has compiled a compendium of solutions to this equation for a variety of surface boundary conditions appropriate to chemical leaching [27].

BOUNDARY CONDITIONS FOR THE TRANSPORT EQUATIONS

Water Transport

Normally when field problems are represented with the water flow equation (A-19), the upper boundary condition is represented by specifying the water evaporation rate. For bare soil conditions, when no crop is present, this is either done explicitly by providing external evaporation data or implicitly through a water evaporation model which shifts between potential and soil-limited conditions. Two examples of such models are given by Ritchie [28] or Tanner and Jury [29]. A recent field study has demonstrated how to field-calibrate completely a soil evaporation model for use in bare soil or partial cover cropping conditions [30].

Evapotranspiration Models

When a crop is present, both the upper boundary condition and the water extraction patterns in the root zone must be obtained.

This is usually done through the aid of an evapotranspiration model. Unless the crop is undergoing stress, water will be lost at a potential rate which is dictated by the external meteorological conditions. In this case, the potential water loss, called the potential evapotranspiration, may be estimated from a knowledge of the appropriate external meteorological variables. The exact variables used to make this estimation depend on the type of model. The most comprehensive model currently in use for the prediction of evapotranspiration is the so-called Penman evapotranspiration equation which requires the following for external inputs: the solar radiation, air temperature, air relative humidity, and wind speed as well as a knowledge of the crop reflectance coefficient or albedo. From this information, an estimate is made of the net radiation and an energy balance is conducted at the canopy surface from which the evapotranspiration is determined as a latent heat deficit [31,32]. In recent years, other models have been proposed which use less information than the Penman equation. Among these are the Priestley and Taylor equation [33] which uses net radiation and air temperature to predict evapotranspiration, the solar radiation correlation which uses air temperature and solar radiation to predict evapotranspiration [34], and an advection modification of the Priestley and Taylor equation [35]. Effectiveness of these different models has been studied under field conditions by Shouse et al. [36].

Perhaps the widest known method for predicting evapotranspiration is the correlation with the evaporation pan [37]. This method is reasonably accurate when applied over large-scale areas the size of watersheds and for time resolutions of the order of months. It is not likely to be very accurate as a daily or weekly gauge of evapotranspiration from a small field. For this reason, methods based on the solar radiation are usually preferred when estimating evapotranspiration.

Water Uptake Functions

As mentioned above, under growing conditions, the water extraction patterns change dramatically for crops as the roots seek out water in new locations. Only under relatively stable conditions such as high frequency irrigation of a fully-developed crop will the water extraction pattern become well-defined as a function of depth. Under such conditions, it has been observed that an exponential depth function adequately describes the extraction patterns for a number of crops [38]. However, when chemical transport to great depth is being considered, details of the water extraction patterns within the root zone become relatively unimportant, and a simple uniform water uptake function may be assumed without great loss of information.

For problems involving leaching processes at the column scale or in a field-plot trial, a water flux rate representing the infiltration rate of water is specified as the upper boundary condition. Or, in the case of ponded water on the soil surface, a constant positive pressure equal to the ponding height is specified. These conditions are used with equation (A-19). In a dynamic simulation of wetting and drying processes, both water input and evaporation rates must be included as part of the upper boundary condition. For this case, the water flux is set equal to the net water input to the surface which is equal to the difference between applied water and evaporated water.

The lower boundary condition for water flow through the unsaturated zone is usually assumed to be the water table if a continuous transport model is represented over the entire unsaturated zone. For simulations representing transport through soil with deep water tables which are below the zone of interest, the lower boundary condition is sometimes represented by assuming gravity flow which means that the gradient of matric potential is set equal to zero at some depth far below the surface.

Chemical Transport

Unless a chemical is actually being added at the soil surface, the normal upper boundary condition for a chemical is contingent on chemical volatilization [3]. Since transport to the atmosphere takes place through the vapor phase, Jury et al. [3], suggested that this boundary condition be written as

$$J_S\,(o,t) = -\frac{D^{AIR}_G}{d}\,[C_G\,(o,t) - C^{AIR}_G\,(d,t)] \qquad \text{(A-36)}$$

where the solute flux J_S is given by equation (A-16), d is the (hypothetical) thickness of the stagnant boundary layer above the soil surface, and C^{AIR}_G is the chemical gas concentration in the well-stirred air above the boundary layer.

In practice, this boundary condition is modified by defining an effective transfer coefficient $h = D^{AIR}_G/d$ when used under field conditions. Jury et al. [3], suggested that the hypothetical boundary layer thickness d could be evaluated by assuming an analogy between water evaporation and chemical volatilization. With this analogy, the measured water evaporation rate could be used to evaluate the boundary layer thickness.

For compounds which volatilize only slightly, the free air concentration C^{AIR}_G may be set equal to zero. In analogy with the simplified version of the transport equation above, one may use the linear equilibrium isotherm and Henry's Law to construct the following form of equation (A-36).

$$-D_E\,\frac{\partial C_T}{\partial z}\,(o,t) + V_E C_T(o,t) = H_E C_T(o,t) \qquad \text{(A-37)}$$

where

$$H_E = D^{AIR}{}_G/dC_G \tag{A-38}$$

is the effective transport coefficient. This approach was used in developing the screening model of Jury et al. [3].

The lower boundary condition for chemicals in soil is usually approximated by assuming no concentration of chemical at great depth. A finite initial concentration incorporated over a layer of variable thickness can be used to make realistic assessments of subsequent movement through soil [3,27].

METHODS OF SOLUTION OF THE TRANSPORT EQUATIONS

Water Flow Equation

Because of the non-linear relationship between hydraulic conductivity and matric potential k(h) and between matric potential and water content h(θ), only numerical solutions are available to the water transport equation (A-19). There are currently two widely used numerical methods of solution to this equation: the finite difference method and the finite element method. The finite difference method appears to be more useful for one-dimensional problems when moderate water fluxes J_W are involved; whereas the finite element method has advantages for complicated multi-dimensional problems [39,40].

Chemical Transport Equation

The chemical transport equation (A-21) when simplified to the convection-dispersion equation given in equation (A-35) may be solved by analytic methods such as Laplace transform or fourier series when the soil and water transport properties are uniform which occurs when the water flux J_W is zero or constant and when the water content θ is constant or nearly constant with depth. This condition is commonly present in miscible displacement column breakthrough studies which have been used to study a large number of chemical and soil physical properties [41]. Analytic solutions have been obtained to equation (A-35) for step function and pulse input boundary conditions and for both zero and first-order chemical reactions occurring in the soil [27]. The combined equations (A-35) and boundary condition (A-37) have been proposed for use in a chemical environmental screening model for large numbers of chemicals whose benchmark properties are known. This model uses an analytic solution based on the Laplace transform [3].

When the chemical transport equation is used in situations in which water or soil transport properties are non-uniform or when transient water flow occurs, a numerical method of solution must be used. As in the case of water flow, this is either a finite difference or a finite element model. For multi-dimensional simulations of chemical transport, the finite element model is usually preferred.

ADSORPTION MODELS

Equilibrium Relations

The relationship between the equilibrium concentration of chemical in solution and the associated chemical adsorbed on the mineral or organic surface is called the adsorption isotherm. Several mathematical relationships have been used to describe this equilibrium relationship in soil water systems. In cases where a maximum adsorbed concentration is reached, the relationship may be fit to a Langmuir isotherm

$$C_S = KC_S^{MAX} C_L/(1 + KC_L) \qquad \text{(A-39)}$$

where C_S is adsorbed concentration (g/g soil), and C_L is solution concentation (g/cm^3 solution/g).

Although the Langmuir relationship has been occasionally used to represent organic chemical adsorption in soil, the Freundlich isotherm, equation (A-40), is much more commonly used to express the adsorption relation

$$C_S = K_f (C_L)^{1/N} \qquad \text{(A-40)}$$

where K_f and N are constant.

Although the Langmuir relationship has a theoretical basis, it applies only to a pure adsorber and adsorbent. Because organic molecules in solution are competing with water molecules for the surface adsorption sites and because the surface adsorbing sites are heterogeneous, the physical basis of the Langmuir isotherm is obscured in a soil-water system. However, the adsorption of organic cations by soil has been shown to fit a Langmuir isotherm [42]. The Freundlich relationship is empirical although it has been related to a special form of the Langmuir relationship in certain cases.

The Freundlich relationship is frequently able to describe adsorption both at low and high chemical concentrations. At dilute concentrations, a linear relationship is frequently used with N = 1

$$C_S = K_D C_L \qquad \text{(A-41)}$$

where K_D is the distribution coefficient. This relationship has considerable advantages for mathematical modeling which are discussed in Chapter 8. Its applicability to soil-water systems has been shown by Karickhoff et al. [18]. However, equation (A-41) should not be thought of as a general relationship for all chemicals. Its limitations are discussed in depth in the review article by Mingelgrin and Gerstl [43].

Rate-Limited Expressions

In a soil-water system where solution is flowing particularly through structured soils, it is unlikely that complete equilibrium is reached between solution and adsorbed concentrations. Although equilibrium is usually assumed in modeling, several researchers have investigated the approach to equilibrium dynamically. Using non-linear kinetics, van Genuchten et al. [20], describe the approach to equilibrium using equation (A-42).

$$\frac{\partial C_s}{\partial t} = K_2 \left[\frac{K_1\theta}{K_2\rho_b} C_L^N - C_s \right] \qquad \text{(A-42)}$$

where K_1 and K_2 are the kinetic rate coefficients for forward and backward reactions, θ is soil volumetric water content, and ρ_b is soil dry bulk density. The quantity in brackets in equation (A-42) is a form of the Freundlich isotherm, equation (A-40). Thus, by assumption, the rate of approach equilibrium is proportional to the deviation from equilibrium. A problem with using an expression like equation (A-42) to describe adsorption is that the rate coefficients K_1, K_2 are not known a priori and must be fitted to column breakthrough data by regression. Since there are also other transport coefficients, such as the dispersion coefficient which must be fitted at the same time, this procedure often masks the physical significance of the parameters obtained in the experiments. Lindstrom et al. [44], proposed a modified model describing the kinetics of adsorption and desorption. Their model differed from that of van Genuchten et al. [20], in that the probability of adsorbing to the surface was allowed to vary with the degree of surface coverage. However, this model required still another parameter which had to be obtained by fitting procedures.

A major complication in describing adsorption of dissolved species in a soil-water system is that all adsorption sites are not equally accessible. Leenheer and Ahlrichs [45], for example, reported that although 60 percent of the adsorption reaction was completed in one minute, adsorption of carbaryl and parathion onto organic matter continued at a slow rate for another two or three hours, and these authors postulated that this was due to the increased time required to reach complete equilibrium with the internal surfaces of organic matter by diffusion. Other complications of rate-limited adsorption modeling are discussed in Rao and Davidson [46].

REFERENCES

1. Bear, J. Dynamics of Fluids in Porous Media (New York: American Elsevier, 1972).

2. Hillel, D. Soil and Water (New York: Academic Press, 1971).

3. Jury, W. A., W. F. Spencer, and W. J. Farmer. "Use of Models for Predicting Relative Volatility, Persistence, and Mobility of Pesticides and Other Trace Organics in Soil Systems," in Hazard Assessment of Chemicals, Vol. 2, J. Saxena, Ed. (New York, Academic Press, 1983).

4. Nielsen, D. R., R. D. Jackson, J. W. Cary, and D. D. Evans. Soil Water (Madison, WI: American Society of Agronomy, 1972).

5. Millington, R. J., and J. M. Quirk. "Permeability of Porous Solids," Trans. Farady Soc. 57:1200-1207 (1961).

6. Letey, J., and W. J. Farmer. "Movement of Pesticides in Soil," in Pesticides in Soil and Water, W. D. Guenzi, Ed. (Madison, WI: Soil Science Society of America, 1974).

7. Shearer, R. C., J. Letey, W. J. Farmer, and A. Klute. "Lindane Diffusion in Soil," Soil Sci. Soc. Amer. Proc. 37:189-194 (1973).

8. Farmer, W. J., M. S. Yang, J. Letey, and W. F. Spencer. "Hexachlorbenzene: Its Vapor Pressure and Vapor Phase Diffusion in Soil," Soil Sci. Soc. Amer. J. 44:676-680 (1980).

9. Jury, W. A., R. Grover, W. F. Spencer, and W. F. Farmer. "Modeling Vapor Losses of Soil-Incorporated Triallate," Soil Sci. Soc. Amer. J. 44:445-450 (1980).

10. Sallam, A., W. A. Jury, and J. Letey. "Measurement of Gas Diffusion Coefficient Under Relatively Low Air-Filled Porosity," Soil Sci. Soc. Amer. J. (In Press).

11. Bresler, E. "Two-Dimensional Transport of Solutes During Nonsteady Infiltration From a Trickle Source," Soil Sci. Soc. Amer. J. 39:604-612 (1975).

12. Kirkham, D., and W. L. Powers. Advanced Soil Physics (New York: Wiley-Interscience, 1972).

13. Call, F. "Soil Fumigation. V. Diffusion of EDB Through Soils," J. Sci. Food Agr. 8:143 (1957).

14. Goring, C. A. I. "Theory and Principles of Soil Fumigation," Adv. Pest Control Res. 5:47 (1962).

15. Leistra, M. "Distribution of 1,3-dichloropropene Over the Phases in Soil," J. Agr. Food Chem. 18:1124 (1970).

16. Spencer, W. F., and M. M. Cliath. "Desorption of Lindane from Soil as Related to Vapor Density," Soil Sci. Soc. Amer. Proc. 34:574-578 (1973).

17. Hamaker, J. W., and J. M. Thompson. "Adsorption," in Organic Chemicals in the Soil Environment, C. A. I. Goring and J. W. Hamaker, Eds. (New York: Marcel-Dekker, Inc., 1972).

18. Karickhoff, S. W., D. S. Brown, and T. A. Scott. "Sorption of Hydrophobic Pollutants on Natural Sediments and Soils," Water Res. 13:241-248 (1979).

19. Karickhoff, S. W. "Semi-Empirical Estimation of Sorption of Hydrophobic Pollutants on Natural Sediments," Chemisphere 10:833-846 (1981).

20. van Genuchten, M. Th., J. M. Davidson, and P. J. Wierenga. "An Evaluation of Kinetic and Equilibrium Equations for the Prediction of Pesticide Movement Through Porous Media," Soil Sci. Soc. Amer. Proc. 38:29-34 (1974).

21. O'Connor, G. A., P. J. Wierenga, H. H. Cheng, and K. G. Doxtader. "Movement of 2,4,5-T Through Large Soil Columns," Soil Sci. 130:157-162 (1980).

22. Lindstrom, F. T., R. Haque, V. H. Freed, and L. Boersma. "Theory on the Movement of Some Herbicides in Soil," Env. Sci. Tech. 1:561-565 (1969).

23. Oddson, J. K., J. Letey, and L. V. Weeks. "Predicted Distribution of Organic Chemicals in Soil," Soil Sci. Soc. Amer. Proc. 34:412-417 (1970).

24. Rao, P. S. C., D. E. Ralston, R. E. Jessup, and J. M. Davidson. "Solute Transport in Aggregated Porous Media," Soil Sci. Soc. Amer. J. 44:1139-1146 (1980).

25. Jury, W. A. "Use of solute transport models to estimate salt balance below irrigated cropland," in Advances in Irrigation, Vol. 1., D. Hillel, Ed. (New York: Academic Press, 1982).

26. Nash, R. G. "Dissipation Rates of Pesticides from Soils," in CREAMS, Vol. 3., W. G. Knisel, Ed. (Washington, DC: U.S. Department of Agriculture, 1980).

27. van Genuchten, M. Th., and W. J. Alves. "Analytical Solutions of the One-Dimensional Convective Dispersive Solute Transport Equation," Technical Bulletin 1661, U.S. Department of Agriculture (1982).

28. Ritchie, J. T. "Model for Predicting Evaporation From a Row Crop With Incomplete Cover," Water Resour. Res. 8:1204-1210 (1972).

29. Tanner, C. B., and W. A. Jury. "Estimating Evaporation and Transpiration from a Row Crop During Incomplete Cover," Agron. J. 68:239-244 (1976).

30. Shouse, P., W. A. Jury, and L. H. Stolzy. "Field Measurement and Modeling of Cowpea Water Use and Yield Under Stressed and Well Watered Growth Conditions," Hilgardia 50:1-25 (1982).

31. Penman, H. L. "Natural Evaporation From Open Water, Bare Soil and Grass," Proc. Roy. Soc. 193:120-146 (1948).

32. Tanner, C. B. "Evaporation of Water from Plants and Soil," in Water Deficits and Plant Growth, T. T. Kozlowski, Ed. (New York: Academic Press, 1968).

33. Priestly, C. H. B., and R. J. Taylor. "On the Assessment of Surface Heat Flux and Evaporation Using Large-Scale Parameters," Monthly Weather Review 100:81-92 (1972).

34. Jensen, M.E., and H. R. Haise. "Estimating Evapotranspiration from Solar Radiation," Amer. Soc. Civ. Eng. Proc. 89:15-41 (1963).

35. Jury, W. A., and C. B. Tanner. "Advection Modification of the Priestly and Taylor Evapotranspiration Formula," Agron. J. 67:840-842 (1975).

36. Shouse, P., W. A. Jury, and L. H. Stolzy. "Use of Deterministic and Empirical Models to Predict Potential Evapotranspiration in a Advective Environment," Agron. J. 72:994-998 (1980).

37. Doorenbos, J., and W. O. Pruitt. "Crop Water Requirements," Irrigation and Drainage Paper 24 (Rome: FAO, 1976).

38. Feddes, R. A., E. Bresler, and S. P. Neuman. "Field Test of a Modified Numerical Model for Water Uptake by Root Systems," Water Resour. Res. 10:1199-1206 (1974).

39. Reeves, M., and J. Duguid. 1975. "Water Movement Through Saturated-Unsaturated Porous Media -- a Finite Element Model." ORNC-4927 (Oak Ridge, TN: Oak Ridge National Laboratory, 1975).

40. van Genuchten, M. Th. "Land Disposal of Hazardous Wastes," EPA-600/9-78-016, Proceedings of the 4th Annual Research Symposium (1978).

41. Bigger, J. W., and D. R. Nielson. "Miscible Displacement and Leaching Phenomenon," Agron. Monog. 11:254-274 (1967).

42. Weber, J. B., P. W. Perry, and R. P. Upchurch. "The Influence of Temperature on the Adsorption of Paraquat, Diquat, 2,4-D, and Prometone by Clays, Charcoal, and an Anion-Exchange Resin," Soil Sci. Soc. Amer. Proc. 29:678-688 (1965).

43. Mingelgrin, U., and Z. Gerstl. "Reevaluation of Partitioning as a Mechanism of Nonionic Chemical Adsorption in Soil," J. Environ. Qual. 12:1-11 (1983).

44. Lindstrom, F. T., R. Haque, and W. R. Coshow. "Adsorption From Solution," J. Phys. Chem. 74:495-502 (1970).

45. Leenheer, J. A., and J. L. Ahlrichs. "A Kinetic and Equilibrium Study of the Adsorption of Carbaryl and Parathion Upon Soil Organic Matter Surfaces," Soil Sci. Soc. Amer. Proc. 35:700-704 (1971).

46. Rao, P. S. C., and J. M. Davidson. "Estimation of Pesticide Retention and Transformation Parameters," in Environmental Impact of Nonpoint Source Pollution, M. R. Overcash and J. M. Davidson, Eds. (Ann Arbor, MI: Ann Arbor Science Publications, 1980).

INDEX